Home VCR Repair Illustrated

Richard C. Wilkins
Cheryl A. Hubbard

DISCARD

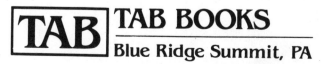

TAB TAB BOOKS
Blue Ridge Summit, PA

FIRST EDITION
FIRST PRINTING

© 1991 by **TAB Books**.
TAB Books is a division of McGraw-Hill, Inc.

Library of Congress Cataloging-in-Publication Data

Wilkins, Richard C.
   Home VCR repair illustrated / by Richard C. Wilkins and Cheryl A.
Hubbard.
     p.   cm.
   Includes index.
   ISBN 0-8306-7711-9 (h)   ISBN 0-8306-3711-7 (p)
   1. Video tape recorders and recording—Maintenance and repair-
Amateurs' manuals. I. Hubbard, Cheryl A. II. Title.
TK9961.W56 1991
621.388'337—dc20   91-12590
                                          CIP

TAB Books offers software for sale. For information and a catalog, please
contact TAB Software Department, Blue Ridge Summit, PA 17294-0850.

Acquisitions Editor: Roland S. Phelps
Bok Editor: John C. Baker
Production: Katherine G. Brown
Book Design: Jaclyn J. Boone   EL1

Contents

Dedication

This book is dedicated to the ones we love.
To Michael, Amber, Harley, and Amanda.
Thank you for all your love and patience and
for giving us of your time and understanding.
This book is for you.

Preface

ONE DAY I WAS SITTING IN MY CAR EATING A CHEESEBURGER. I WAS DWELLING on an idea of how to teach individuals to repair VCRs themselves. My job had run out of excitement. Everyday was fixing the same simple problems, with no challenges. I decided to give myself the biggest challenge of my life. I decided the answer was making a video tape. About halfway through the process of making the tape, I realized there wasn't any way to get that much information in a two-hour video. The stopper was the response I kept receiving when I delivered my idea. Everyone was asking me, "If my VCR doesn't work, how am I supposed to watch your video on fixing one?" That question inspired me to write this book.

I was traveling at the time, working around the country. I came to the realization that the majority of the VCR repair shops were scamming the unknowing public. The goal I have for this book is to teach anyone how to avoid spending $100 for a $10 job. There isn't one problem with a VCR covered in this book that costs more than $15 to repair (tools and supplies excluded). The same problems, if they are taken in for servicing, are going to cost between $80 and $130. At a supposedly reputable shop, a maintenance kit and having the video head cleaned and the dust blown out ran the consumer about $110.

The icing on this book is that I've included every trade secret I could get my hands on. I've labeled only a few, but there are many throughout the book. Another plus is that I've eliminated all testing equipment, showing you how to use common household items to assist you. All technical aspects of electronics are presented through my experiences in repairing VCRs since they were first introduced to the consumer. When using a manual, test equipment, and technical theories, a roller guide alignment takes about two hours on the average. Using my method takes about 15 minutes, without any testing equipment. There's no difference in accuracy from their method to mine, just a lot of time and money.

Acknowledgments

THANK YOU TO GENE STOTTS, WALTER DALLIS, AND ROGER HARRIS WHO GAVE of their expertise, knowledge, and guidance.

Thank you to Robert Pesavento, Dawn Robinette, Cindy Knox, and Anita Turner who assisted in the preparation and production of the beginning manuscripts.

To our newest friends, Chris Rowe and Shiloh Maches of Video Brite Associates. Thank you for all the time given in the final preparation of the manuscript.

To Mel Martin and Stephan Wright, a special thanks for your true blessing.

To Barbara Brooks Siegle, also known as "Mom," thank you for the contribution of your time and talent on the artwork.

To Denny Thompson, the closest friend a man can have. Thank you for all your moral support.

A very special thanks to Wenda Sudduth and Linda Barum for the extra effort given in artwork and word processing production. Thanks for understanding as only close friends can, the importance this book meant to me. For sacrificing time and giving of yourselves.

A special thanks to TAB Books, especially Roland Phelps for guiding us through the path of writing and producing a manuscript for publication.

To Cheryl Hubbard, my special Cherrie. Thank you for the days, nights, and hours of time given to polish and prepare the final manuscript. For the mental and spiritual support given on our breaks. I couldn't have asked for a more dedicated co-author.

Especially to Johnathan Siegle, only he knows the reasons why.

Introduction

THIS BOOK WAS WRITTEN BASED UPON MY IDEAS AND EXPERIENCE IN VCR repair.

The method is simple and safe to use on any VCR, if you follow the directions carefully. It's designed to teach you how to do-it-yourself in your own home, using household items and basic tools. No test equipment is involved. This book gives you step-by-step directions.

You will probably find it helpful to refer to chapter 14 to diagnose the problem in your VCR. This book is broken down into sections that correspond to each particular section of a VCR, so you can proceed to the appropriate section.

Most chapters contain a review section. This section provides step-by-step instructions to assist in diagnosing and repairing the VCR. This section also acts as a checklist, so you don't miss any steps when performing repairs.

Each chapter contains referrals for further information in other chapters, to give you the best understanding possible. This book zeros in on a problem, locates it, and shows how to remove, repair, and replace any part. Besides providing information on repair, this book contains knowledge previously not accessible to everyone.

1
Tools and supplies needed

IF YOU USE A VCR AS MUCH AS I DO, YOU KNOW HOW FRUSTRATING IT CAN BE to watch a distorted picture or no picture at all. To top it off, you enjoy and take pride in fixing things yourself, and you would fix your VCR if only you knew how. All it takes to save a lot of time and money is some basic information. After knowing how relatively simple and inexpensive it is to do repairs yourself, you'll probably wish this book was available a long time ago.

Eventually a VCR requires cleaning and lubricating, especially if you view rented video cassettes. There's no reason to pay a service charge when routine cleaning and maintenance are required. Just a little preventative maintenance saves you money and prevents costly repairs in the future.

Tools

The tools you need to begin are a small-tip magnetic Phillips screwdriver and two sturdy flathead screwdrivers, one standard size tip and one very small. If you don't happen to have a magnetic Phillips screwdriver, don't purchase one. I'll show you how to magnetize a Phillips head screwdriver. You also will need to have a soft bristled paintbrush, 1.5mm, 1.27mm, and 0.89mm allen wrenches, long-nose pliers, wirecutters, $\frac{7}{32}$-inch socket or $\frac{7}{32}$-inch nutdriver, an alignment screwdriver, a tube of super glue, and clear silicone rubber sealer. These items can be purchased at a local hardware store.

Supplies

Because you're just starting out, here's an important shopping list: one package of chamois sticks, head cleaner or cleaning alcohol, a tube of phono lube, a container of tuner grease, and a spray can of degreaser. The degreaser must be the type that'll completely evaporate. Check for complete evaporation by spraying it on your fingers and rubbing them together; the spray should leave no film or residue. You'll also need 24-gauge stranded wire, solder

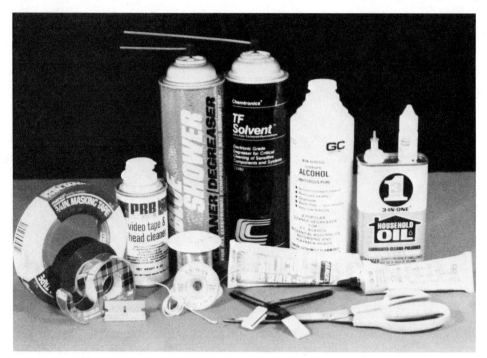

1-1 Chemicals and supplies needed.

wick, rosin type solder, a solder iron or gun, and a C-ring puller. These items can be purchased at a local electronics supply store. See Fig. 1-1 for examples.

Have on hand a roll of black electrical tape, masking tape, ¾-inch scotch tape, a single-edge razor blade, an old toothbrush, a sharp pair of scissors, paper towels, a can of nonspraying household oil, two nutdigging tools or two cuticle-remover nail-cleaning tools, and a clean dry empty lid or container. These items can be purchased at a local drug store. Last, but not least, is a TV set. The television acts as a monitor, helping you diagnose problem areas in a VCR.

Making a glass brush

A glass brush is useful and simplifies repairs. A glass brush isn't ready made. To make one, you'll need an alignment screwdriver, as shown in Fig. 1-2. The screwdriver shaft is made of fiberglass. To transform it into a fiberglass wand, first cut off the shaft flush at the handle. Next, heat an electric stove or hot plate to medium heat. Don't let the element get red hot. Place the cut off end of the wand onto the element. Push down the wand while twisting it back and forth until a white brush appears, as shown in Fig. 1-3. If an electric stove or hot plate isn't available, you can use a hot solder iron to produce the white brush (Fig. 1-4). You've just made a pure glass brush. A glass brush is useful as a cleaning utensil.

Magnetizing a screwdriver

Find a magnet; any magnet will do. Rub the tip of the screwdriver across the magnet. The screwdriver now is magnetized for a short period of time. When magnetized, a screwdriver allows you to replace and retrieve those hard-to-get-at screws.

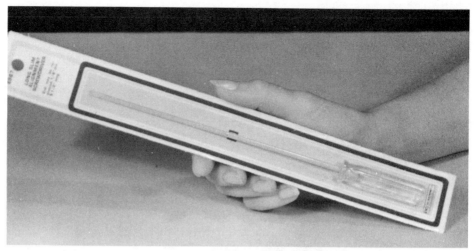

1-2 An alignment screwdriver.

1-3 Making a glass brush.

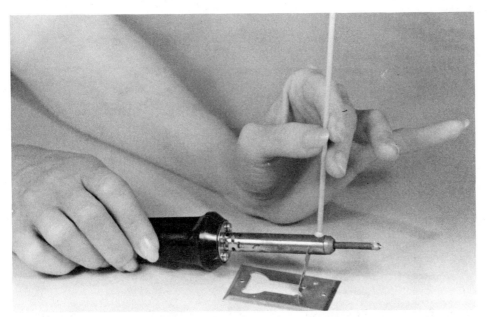

1-4 Using a soldering iron to make a glass brush.

2
Getting inside a VCR

BEFORE ANY REPAIRS CAN BE DONE ON YOUR VCR, YOU MUST FIRST OPEN IT, so you can get to the source of the problem. For the first-time VCR handyman, this part of the repair process can be the most intimidating. Each time you repair your VCR, you will gain confidence, and soon you'll be disassembling your VCR like an experienced repairman.

Seven ways to remove the main top cover

The main top cover must be removed to get inside a VCR. There are seven different ways a top cover might be mounted. In the majority of all models a Phillips or flathead screwdriver is the only tool required for removal.

First way

In Fig. 2-1, the top cover screws are on the top rear portion of the VCR. After removing the screws, lift up the back portion of the cover and pull straight back. This releases the front latches, as shown in Fig. 2-2. The cover now lifts right off. Put the screws in a safe place.

Second way

Locate the main top cover screws near the back of the VCR, as shown in Fig. 2-3. Look at the cassette lid, you might find two screws on top (Fig. 2-4). The cassette lid has to come off to remove the top cover. After removing the cassette lid screws, push Eject to pop up the cassette holder. With some models, the unit must be plugged in for Eject to function. Pick up on the back of the cassette lid and push the lid forward (Fig. 2-5) to release the front latches on the lid for removal. After removing the cassette lid, push the cassette holder down to its locked position. Remove all the screws and grasp the rear of the cover at the bottom and pull it straight up four to six inches. Then pull straight back to release the front latches, as shown in Fig. 2-6. In this model, the top, sides, and back come off in one piece.

2-1   First location of cover screws.

2-2   Releasing the front latches.

2-3 Second location of cover screws.

2-4 Locating the cassette lid screws.

2-5 Removing the cassette lid.

2-6 Releasing the front latches.

2-7 The hidden top cover screw.

In a few models, there might be one more screw to remove. Look to see if the unit has a door on the top cover. Usually there are fine tuning dials under this door. There might be a hidden top cover screw (Fig. 2-7). One model has a hidden screw on the lip of the top cover under the cassette lid. Remove the cassette lid to find the hidden screw.

Third way

Some models don't have any screws on top of the cassette lid. If not, don't remove the lid. To take the top cover off, push eject to pop up the cassette holder. Next, remove the top cover screws. Pick up the back of the cover and pull it straight back to release the front lip. Then slide the top cover forward over the cassette holder for removal, as shown in Fig. 2-8.

Fourth way

In a few older models, a lip across the back of the top cover prohibits lifting the back of the cover for removal, even after you have removed all of the top cover screws. To remove this type of cover, loosen the screws on the back cover about a ½ inch. Pull the back cover out to clear the lip, as shown in Fig. 2-9. Remember the top cover should come off easily. Don't force it. If it doesn't come right off, there's another screw to remove.

Fifth way

Usually on frontload models (Fig. 2-10), the top cover screws are on both sides of the unit. In some models, an additional screw is in the middle of the back cover. In other models, the

2-8   Clearing the cassette lid for removal.

2-9   Clearing the back cover lip.

screws are located across the back with no screws on the sides. These screws fasten the top cover to the back cover. Remove the top cover screws, and grab hold of the cover on the bottom of each side, lifting up then pulling straight back. This procedure releases the front lip, as shown in Fig. 2-11. Now, the cover lifts straight up and off. In these models, the top and sides come off in one piece.

2-10   Third location of cover screws.

2-11   Releasing the front lip.

Sixth way

In some frontload models, the cover still won't come off, even after following the above procedures. Place the unit on its side, then look at the screws on the bottom cover plate. See if it has two to three screws with arrows pointed at each one. Usually there's one screw

2-12   Arrows pointing at the cover screws.

on each side, or one on one side and two on the other, as shown in Fig. 2-12. Remove the two or three screws, and place the unit back right side up. Pull the cover back slightly, and lift the cover off. If you find arrows pointed at all the screws on the bottom cover plate, don't remove them. These arrows pertain only to the bottom cover.

Seventh way

After removing the screws, slide the cover straight back to release the latches. There's usually two or four hidden L-shaped latches on the inside of the main cover holding the top cover down. You can determine if you have this type of unit, by the location of the mounting screws. These screws are inserted through the back lip on the main top cover. In some models, you might find two more screws with arrows pointing at them on the bottom cover. Remove these screws, slide the top cover straight back about a ½ inch, and pick up the sides, pulling it straight up and off. After removing the top cover, you can see the hidden latches, as shown in Fig. 2-13. When remounting the top cover, align the latches on the chassis with the slots on the bottom lip of the cover. Slide the cover forward to secure the latches and replace the screws.

Electrical shock

Now that you've removed the main top cover, I must discuss avoiding an electrical shock. Every VCR has a manufacturer's warning label that cautions you about electrical shock. I

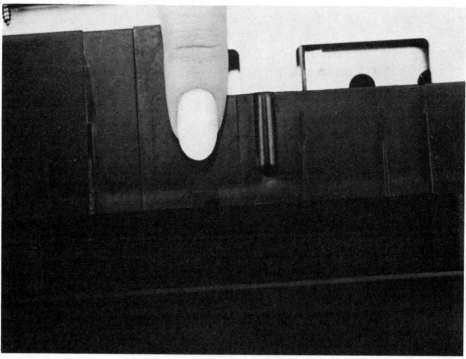

2-13 Hidden top cover latches.

want you to feel comfortable working on your VCR. You can touch the electrical circuits while the unit is on without getting shocked (Fig. 2-14). All units run on a low dc voltage, ranging from 5 to 30 volts. This voltage isn't harmful.

The only location that can harm you is the power supply. The power supply converts the 110 volts into a low voltage. To locate the power supply, follow the power cord through the back of the unit and into a covered connection box. Inside the box is the main fuse. The connection box is usually encased with plastic or metal, and it's grounded to the chassis to protect you. Do not, under any circumstances, remove this protective cover while the unit is plugged in. From this box, the 110 volts goes directly into the power transformer, as shown in Fig. 2-15. The purpose of the power transformer is to reduce the 110 voltage to the variable low source that makes the VCR function.

In a few models, the 110 volts is exposed (Fig. 2-16). Follow the power cord through the back of the unit to where it connects to a circuit board. Beside this connection on the circuit board, you'll find the main 110-volt power fuse. Behind the power fuse, there'll be a 110-volt outlet. These are the only two points where you can get shocked. From there the power goes directly into the power transformer. If you find an exposed 110-volt connection, unplug the unit and place black electrical tape over the exposed area. The tape protects you against electrical shock, as shown in Fig. 2-17. Another precaution is to take a thin sheet of plastic or cardboard and place it over and around the entire exposed area. Then use black electrical tape to hold it down.

In some newer models, the VCR has no power transformer. The voltage is reduced electronically on the power supply board. To locate this board, follow the power cord into the

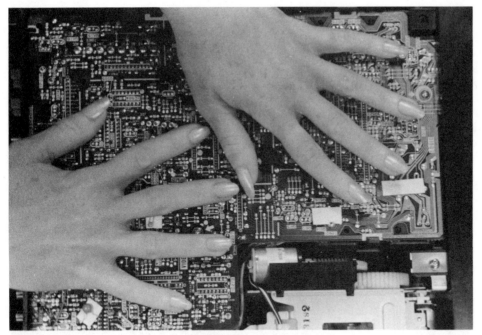

2-14 Touching the electrical circuits.

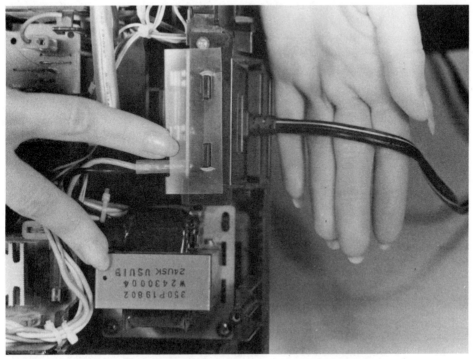

2-15 A covered box (top) and a power transformer (bottom).

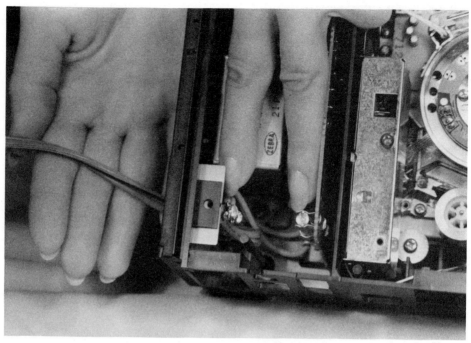

2-16 Exposed 110-volt connections.

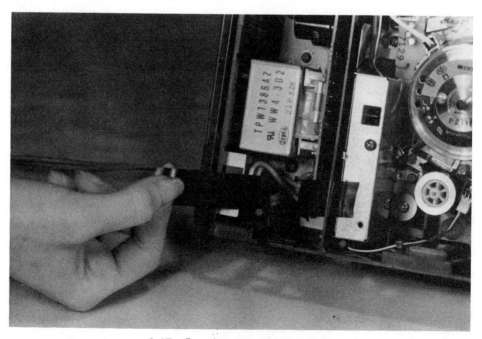

2-17 Covering exposed connections.

unit and follow it to the power supply board. Some of these boards are covered with metal all the way around it. If the unit has an exposed power board, you'll need to cover the board with soft plastic or with cardboard. You can do this by placing a flat piece of cardboard over the top of the board in the same position where the top cover was located. Secure the cardboard down with black electrical tape to hold it into place.

Working conditions to avoid

Working on a VCR on or around an electric stove or any other electrical apparatus.
Working close to any sinks made of metal. Standing barefoot on a cement floor.
Standing in a puddle of water.
Wearing rings while working on a VCR.
Soldering a part while the unit is plugged in.

The VCR should be plugged in only when

You are diagnosing the problems in the VCR by watching the lines on the TV monitor.
The VCR is in play, fast forward, or rewind and you're looking for a mechinal problem.
Checking a cassette lamp.
Checking the audio amplifier.
Checking the video head amplifier.
Doing a audio or video alignment.
Making the torque adjustment.
Centering the tracking control.
Checking for a bad dc motor.

Under all other circumstances, the VCR should be unplugged before you begin repairs.

Review

☐ Always make sure you have the proper tools to work with.

☐ If the top cover doesn't come off easily, look for another screw.

☐ Only remove cassette holder lids that have screws on top of the lid.

☐ Cover any exposed 110-volt terminals with black electrical tape to protect you.

☐ Never stick a screwdriver in the covered box.

☐ Never lay tools on top of a circuit board.

☐ Never grab hold of an electrical appliance while working on a VCR.

☐ Do not wear rings while working inside the VCR.

☐ These precautions are here to protect you. Follow them carefully.

I've discussed many types of top covers and methods for removing them. The next chapter covers the most crucial portion of a VCR: the path the video tape travels.

3

Locating the video tape path

BEFORE YOU CAN LEARN HOW TO CLEAN AND SERVICE A VCR, FIRST YOU MUST learn the video tape path. The video tape path is the most crucial portion of a VCR. There are hundreds of different makes and models of VCRs; however, from the very first model produced to the most current on the market today, the video tape path remains the same. Study Fig. 3-1, then I'll explain the video tape path.

How the tape path works

When a tape is loaded, the video tape path begins, starting from the left-hand side of the video cassette. As the tape comes off the supply reel, its first contact is the supply tape guide inside the video cassette. The supply tape guide secures the video tape on its proper path. The tape then passes over the back tension guide, which maintains the right amount of torque. The tape travels across a supply tape guide on the transport, the chassis of the carriage that holds the carriage together. The supply tape guide secures the video tape on its proper path. The tape then passes the erase head, which erases previously recorded information when in the record mode. Next, the tape travels up against the tension roller. The tension roller prevents the tape from shuttering and keeps the tape running smooth. The tape travels around the supply roller guide, aligning the video tape to the left-side of the video drum. The video drum houses the video heads, which read the signals off the video tape. This process enables you to see images on the TV screen.

The video tape continues around the video drum and travels to the take-up roller guide. The take-up roller guide aligns the video tape to the right side of the video drum. In some models, the tape passes up against another tension roller. It then travels to the A/C head. A/C stands for audio and control. The A/C head function is to play and record the audio portion of the image. The tape crosses the take-up tape guide, securing the video tape on its proper path. It then travels between the capstan shaft and pinch roller. The capstan shaft turns and pulls the video tape through the tape path. The pinch roller clamps up against the

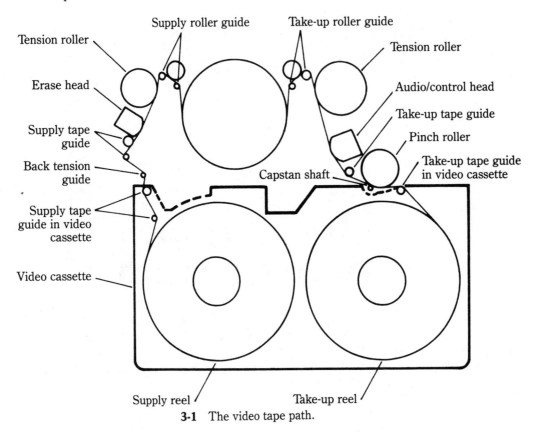

Tension roller

Supply roller guide

Take-up roller guide

Tension roller

Erase head

Audio/control head

Take-up tape guide

Supply tape guide

Pinch roller

Back tension guide

Capstan shaft

Take-up tape guide in video cassette

Supply tape guide in video cassette

Video cassette

Supply reel

Take-up reel

3-1 The video tape path.

capstan shaft to provide the pressure necessary for the capstan shaft to pull the tape. The tape continues to another take-up guide in the video cassette. This guide aligns the video tape to the take-up reel, the spool the video tape rewinds on inside the video cassette.

Removing shield plates covering the tape path

A shield plate must be removed to service and clean the video tape path. The shield plate (Fig. 3-2) is a metal cover that keeps dust off the video heads and prevents electrical interference.

Unplug the VCR before removing the shield plate and locate the screws holding the shield on. Sometimes there'll be one mounting screw, but there can be up to six. There also might be a wire attached to some of the shield plate screws, as shown in Fig. 3-2. You might find two wires attached to two plate screws, or just one wire attached to one plate screw. In most cases, there'll be no wires attached. These attached wires are called grounding wires. A grounding wire grounds the shield plate. It also can be used to ground other parts of the VCR. If a wire is attached to a plate screw, remove it by removing the screw, pulling the wire clip off, and tucking the wire or wires out of the way. In Fig. 3-3, see how the wires have been tucked away to keep them from shorting out other circuits.

Check to see if you've found all the screws. On certain models, you might find a shield plate screw mounted to the transport, as shown in Fig. 3-4. While reaching down to retrieve

3-2 Shield plate and grounding wires.

the shield, it's possible to drop the screw down inside the unit and lose it. If this happens, a magnetic screwdriver will come in handy.

After removing the mounting screws, the next step is to remove the shield plate. Grab the shield plate and pull it straight up and out. Figure 3-5 shows the video tape path, which can be seen after removing the shield plate.

Information on screws

When removing the screws, notice the type of thread on each screw. Screws have different threads. If you place a screw in the wrong screw hole, you can strip the threads. Fine thread screws go into a metal casing. Course screws go into a plastic casing. This is the general rule, the exception to this rule is top shield plate cover.

If you accidently drop a screw inside the unit, retrieve it by turning the unit upside down and shaking it, or take the unit apart. To avoid this inconvenience, use a magnetic screwdriver. When replacing a screw, the magnetic field holds it secure to the screwdriver.

Other types of shield plates

Some shield plates have no grounding wires to be removed. On this make of shield plate, you don't have to completely remove the screws. Just loosen the screws, slide the shield plate to the left, then pull the shield plate straight up and off, as shown in Fig. 3-6.

Another type of shield plate has screws, and clips holding down the cover plate. After removing the screws, you'll find two black clips in each corner on the backside of the plate.

3-3 Tucking away the grounding wires.

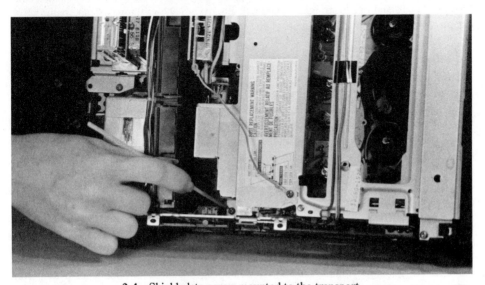

3-4 Shield plate screw mounted to the transport.

3-5 Removing the shield plate.

3-6 A removed shield plate.

Take hold of the clip and push it back, lifting up on the plate at the same time. Do both clips in the same manner.

Some shield plates have an extension going up one or both sides of the cassette carriage (Fig. 16-14). You'll find another mounting screw at the end of this extension.

Removing circuit boards blocking the tape path

There are several ways to remove the circuit board in your VCR. Several of these are detailed below. One of them will be appropriate for your particular model.

First way

In some frontload models, a circuit board is placed over the top of the tape path, and is usually hinged. The hinges are usually at the back of the board as you face the unit. Mounting screws hold it down at the front of this circuit board, which is mounted to the cassette carriage. The cassette carriage is the complete assembly or the housing for the cassette holder. The cassette holder is the compartment you insert the video tape into. You'll find two or three mounting screws on this board, as shown in Fig. 3-7. Remove the mounting screws.

In some models, you'll find a grounding leaf spring. The leaf spring grounds the main top cover to the chassis. This spring has to be removed to pull up the circuit board. Most grounding springs are mounted with a Phillips-head screw (Fig. 3-8). Take out the screw to remove the grounding wire. Take hold of the front of the circuit board and pull straight up. The board opens like a door exposing the shield plate. In Fig. 3-9, this model of shield plate has two mounting screws and no grounding wires. Remove the screws and lift off the shield plate.

Second way

In other models with circuit boards covering the tape path, you'll find one to three screws on the board to be removed. These screws are usually red in color and have arrows or circles around them printed on the circuit board. Look for one to four black latches around the outside of the circuit board, or protruding through a small rectangular hole in the middle of the circuit board. To release these latches, pull straight back away from the board simultaneously lifting the circuit board, as shown in Fig. 3-10. Release each latch until the circuit board opens like a door.

Third way

In some models, there are no hinges. What you'll find is four to six mounting screws. These screws are marked with an arrow or have a circle around them. The screws might be red in color. Usually there's two screws on the cassette carriage, two screws on the chassis, and two screws go directly into the back of the unit. Remove all six screws and lift the circuit board straight up. In this model, part of the back comes out with it, containing the in and out video plugs, as shown in Fig. 3-11. As you remove the board, flip it over the back of the unit onto the table. In some models, the two back screws are on the part containing the in and out video plugs. Also in this type of model, there might be a screw directly beneath the in and out video plugs. For additional information on removing circuit boards, refer to chapter 36 and read the section "Boards underneath the top cover" under the subhead "Removing circuit boards."

3-7   Locating the mounting screws on a circuit board.

3-8   A grounding leaf spring.

3-9 A shield plate and open circuit board.

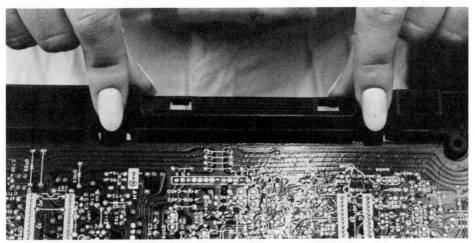

3-10 Releasing the latches.

3-11 Removing a circuit board.

Holding the circuit board open

To keep a circuit board with hinges open for cleaning and repairs, place any nonmetallic object, like a glass brush, between the top of the chassis to the right side of the cassette carriage and the bottom of the circuit board.

Review

☐ The tape path is the same in all models of VCRs.

☐ In some recent models, the tape path is covered by a shield plate that has to be removed for servicing.

☐ Remember to reattach any grounding wires to the shield plate mounting screws.

☐ Use a magnetic screwdriver for those hard-to-get-at screws.

☐ All screws have different threads, so when replacing the screws, make sure you put the right screw in the right screw hole.

☐ Most units with circuit boards are hinged and open like a door.

☐ Remove the mounting screws marked with red screws, arrows, or circles around them.

☐ In some models, remove the circuit board with part of the back and place it over the back of the unit.

The next chapter is on the cleaning process. If good maintenance is followed, a VCR functions more efficiently and generally has fewer mechanical malfunctions.

4
The cleaning process

IT'S IMPORTANT TO BE PREPARED BY HAVING ALL NECESSARY TOOLS AND materials. Make sure to gather everything needed before beginning the cleaning process.

List of items needed

- ☐ One can of degreaser
- ☐ A roll of paper towels
- ☐ Cleaning alcohol or head cleaner
- ☐ Chamois sticks
- ☐ Glass brush
- ☐ Soft-bristle paint brush
- ☐ Empty lid or a small container

Removing the dust and dirt

Unplug the VCR and tip the unit on its side, with the cassette carriage side down. If the unit has accumulated a large amount of lint and dust, then first blow it out with an air compressor or vacuum cleaner. Taking a paint brush, start dusting off your VCR at the top of the unit where the circuit boards are and work your way down to the bottom of the unit, as shown in Fig. 4-1. By dusting it off in this manner, the dirt falls onto the table, not down inside the unit. When you reach the video drum, be careful to brush only the top of the drum. Do not brush the sides of the drum. Brushing the sides can cause damage to the video heads. Make sure to get down inside all the nooks and crannies. VCRs collect a lot of dust. If you have a topload unit, push Eject to pop up the cassette holder. Then brush off the parts underneath the cassette holder. Remember most units require electricity to activate the eject mode.

There are some areas the brush is unable to reach. You also might find dust and oil have accumulated, causing a sticky film. The brush will only smear the film, not remove it (Fig. 4-2). Remedy this by taking a can of degreaser, and spraying from the top of the unit to the

4-1 Dusting a VCR.

bottom of the cassette carriage. This rinses all the residue towards bottom end of the unit for removal.

Make sure to spray into all those hard-to-reach areas by using the nozzle extender, as shown in Fig. 4-3. Spray liberally; degreaser won't damage any part of a VCR. Take a papertowel and collect all the dirt accumulated at the bottom of the unit, as shown in Fig. 4-4.

I've found instances in which a silicone spray has been used to clean or lubricate. If this spray is used, a VCR can completely stop functioning. Silicone spray leaves a film residue, which causes the belts and pulleys to slip. If this has happened to you, just respray it with a can of degreaser. The degreaser will remove all the silicone spray. After you're satisfied the chassis is clean, lay the unit flat down on the table.

Cleaning the various components

After you have given the unit a general cleaning, there are certain parts that you should give special care to clean. These parts and the methods to clean them are described in this sec-

4-2 A sticky film caused by dust and oil.

tion. Now, to clean the video tape path, pour head cleaner or cleaning alcohol into a small container. Take a chamois stick, and immerse it into the solution.

Capstan shaft

There'll be a black residue caked on the surface of the capstan shaft. This residue needs to be removed. Take a saturated chamois stick and rub it up and down the shaft until the exposed portion of the shaft is clean. (See Fig. 4-5. I've removed the cassette holder, to give you a better view.)

In most VCRs, the front left side of the capstan shaft is the only area you can reach. After cleaning this area, the next step is to rotate the shaft. First try turning the shaft. If you have a direct drive model, the shaft turns easily. With some models, the capstan shaft turns with more difficulty, but does turn. If it won't turn, you have a belt driven unit. Look for a capstan motor flywheel, as shown in Fig. 4-6. If the unit has an exposed flywheel, place a finger on top of it and spin it in either direction. This procedure easily turns the capstan shaft for cleaning. If you don't have an exposed flywheel, use the following procedure. Plug in the unit, insert a video tape and push Play and then stop. Eject the tape. The capstan shaft stops in a new position. Now you can clean this portion of the shaft. You might need to repeat this

4-3 Removing the sticky film by spraying it with degreaser.

step two or three times to clean all the way around the shaft. When the capstan shaft is completely clean, it'll be nice and shiny. In older units the capstan shaft is easily accessible and you can clean all the way around it without turning the shaft.

Sometimes this black residue is quite hard and a chamois stick can't remove it. To remove this residue take a glass brush and clean it in the same manner.

To use a glass brush, fill a small container with cleaning alcohol or head cleaner. Dip the brush into the solution, then with a scrubbing motion remove any residue that's difficult to get off. When the brush turns black with residue, stop and clean it off by turning it back and forth in a papertowel. Then redip the brush into the solution, repeating this procedure until the part is clean.

Pinch roller

Place your finger on top of the pinch roller. Because the pinch roller moves freely, a finger will stabilize it (Fig. 4-7). Take a saturated chamois stick, rub it up and down the roller until it's clean. Using your finger, rotate it slowly until you've cleaned all the way around it. Notice there'll be a lot of black residue coming off the pinch roller. You might need to use two or three different chamois sticks to completely clean the pinch roller.

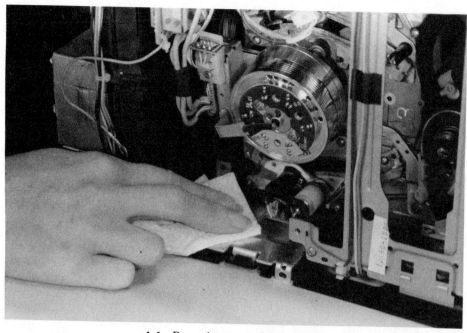

4-4 Removing accumulated residue.

4-5 Cleaning the capstan shaft.

4-6   The capstan motor flywheel.

4-7   Cleaning and stabilizing a pinch roller.

4-8 Cleaning the A/C head.

A/C head

Place a clean chamois stick horizontally across the A/C head and rub it back and forth until the head is clean, as shown in Fig. 4-8. Sometimes you'll find hard black residual buildup on the A/C head. This residue is hard to get off. Just keep scrubbing until it's removed, or you can use a glass brush to remove it. If you're using a glass brush, use a chamois stick to clean any fibers left by the brush. Do not try using a screwdriver or a hard object to remove these hard spots. You'll damage the A/C heads.

Tension rollers and erase head

The next part to clean is the tension roller. Some models contain one tension roller; others contain two. You clean the tension rollers in the same manner as the pinch roller. As you follow the video tape path around, the next part you will come across is the video drum. Leave it for now; I'll show you that procedure in chapter 5. Just continue on to the next tension roller. Locate the eraser head and clean it in the same manner as you cleaned the A/C head.

4-9   Cleaning the roller guides.

Roller guides

Place a chamois stick in a horizontal position onto the glass portion of the roller guides. Scrub up and down to remove any residue. Be sure to go all the way around each roller guide, as shown in Fig. 4-9.

Review

☐ Tip the VCR on its side with the cassette carriage side down.

☐ Brush off all dust, starting at the top of the unit and working down to the bottom.

☐ If necessary, spray with degreaser.

☐ Lay the unit flat on a table, then clean the video tape path using a saturated chamois stick.

☐ Make sure to clean:
  - Capstan shaft
  - Pinch roller
  - A/C head
  - Tension rollers
  - Eraser head
  - Roller guides

Making a checklist of parts will help you remember what parts you've cleaned, and what parts still need to be cleaned.

5

Cleaning the video heads

ALL VIDEO HEADS APPEAR AND FUNCTION THE SAME. VIDEO HEADS ARE located inside the video drum. I've removed a video drum from a VCR in order to demonstrate its functions.

The video head

A video drum consists of two parts: the upper video drum that rotates, and the lower video drum that is stationary and is bolted to the transport. As you observe the upper video drum, you'll find a little slot on each side, directly across from each other, as shown in Fig. 5-1. These slots contain the video heads. In each slot you'll find the protruding tip of a video head. This protruding tip is what you need to clean.

I've disassembled the upper video drum to give a better view of the video heads (Fig. 5-2). This model has two heads. In some models, you might find three or four video heads, or slots, on the side of the video drum. When working on a VCR, *do not, under any circumstances, remove the video drum.* The video drum is a delicate and expensive part.

Why video heads stop working

Look at a close-up view of a video head, as shown in Fig. 5-3. The video heads are made of a metallic carbon compound, which is extremely fragile and expensive. Notice the viewport, or gap, of the video head. This portion reads the signal off the video tape. One reason a video head malfunctions is that the viewport becomes obstructed with dust or lint from dirty tapes. Consequently, the head cannot read the signal. The video heads are quite active, the heart beat of the VCR; keeping them clean is extremely important.

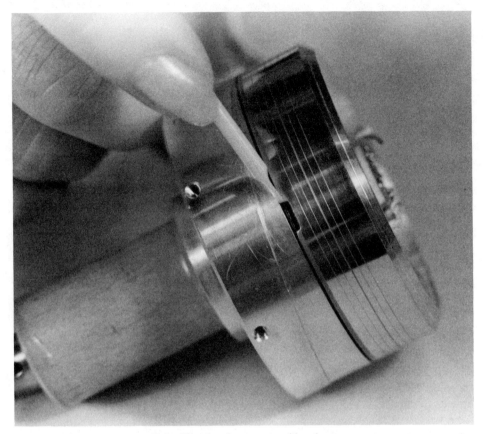

5-1   The slots in an upper video drum.

Cleaning the video heads

Pour some head cleaner or cleaning alcohol into a small container. Take a new chamois stick and saturate it in the cleaner. Apply the chamois stick (Fig. 5-4) in a flat position, so half the chamois is on the lower drum and half is on the upper drum. Hold the stick at a slight angle. Place your finger on top of the video drum. Start rotating the upper drum in a counterclockwise direction, rotating one revolution every four seconds. As you slowly rotate the upper drum, move the chamois stick up and off the drum in one direction, towards the top of the video drum, as shown in Fig. 5-5. Do not pull the chamois stick away from the video drum before you reach the top. Let the chamois stick go all the way up and off. This way you won't leave any dirt deposits on the drum. Repeat this procedure two or three times. *Using a chamois stick in an up and down motion causes damage to the heads.* Do not use this position (Fig. 5-6).

In a few models a rotary cap is on top of the video drum. The rotary cap is part of the drum motor that makes the drum spin. In this case, place your finger on the top corner edge to rotate the drum, as shown in Fig. 5-7. Clean the video drum as previously explained. Avoid using cotton swabs, their fibers get lodged inside the view port, making the head dirtier.

5-2 The video heads.

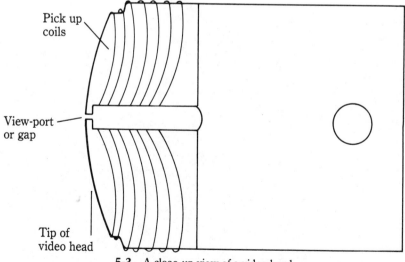

Pick up
coils

View-port
or gap

Tip of
video head

5-3 A close-up view of a video head.

5-4   Applying a chamois stick at the correct angle and rotating the upper video drum.

5-5   Moving the chamois stick up and off the video drum.

5-6 An incorrect position for a chamois stick.

5-7 Rotating a video drum with a rotary cap.

Head cleaning tapes

Cleaning tapes don't always work. Some cleaning tapes just smear the dirt and film leaving a trail of dirt deposits in a ring around the upper video drum. Smearing the dirt deposits can clog the viewport of the video head even more, as shown in Fig. 5-8. The construction of the video head creates a need for a certain amount of pressure to be applied in order to remove the dirt from the viewport. This pressure is applied when cleaning the video drum by hand. Keep the chamois stick flat against the video drum at a slight angle and apply a steady amount of pressure as you move the chamois stick up and off the drum. Now that the video heads and the entire unit are clean, it's time to hook up the VCR to the TV monitor and learn how it functions.

Connecting the VCR to a TV monitor

Attach a coax cable to the out plug of the VCR, and the other end to the in plug on the TV. Attach the antenna, or outside cable, to the in plug on the VCR. Plug in both units and turn them on. Select the proper channel, 3 or 4, on the TV. The VCR/TV switch must be placed in the VCR position. Select a channel on the VCR showing a broadcasted picture on the TV screen. This shows the VCR is producing the signal, not the TV. If both units are connected and fine-tuned properly, you should have a picture of what is being shown on that channel. By following this procedure, you've turned the TV into a monitor, a useful tool for diagnosing problems in a VCR.

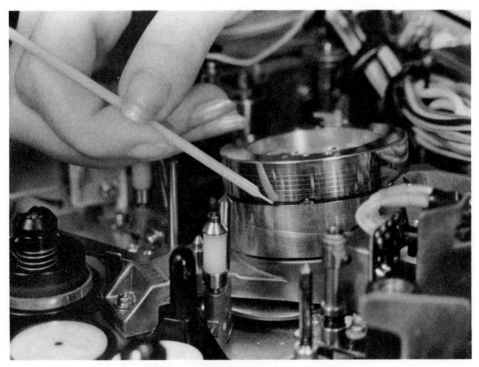

5-8   Residue ring around the upper video drum.

Dirty video heads

Some VCRs have two video heads, while others have four. The symptoms of dirty heads is different for each, as are the steps necessary to get the heads working properly again. See the section that pertains to your VCR for cleaning instructions.

A two-headed machine

Now that everything is ready, insert a video tape and push Play. The VCR should be producing a good picture. Now I'm going to tell you the symptoms of a dirty video head and describe what it'll look like on a TV monitor. If the heads are partially clogged, you'll receive a partially distorted picture, as shown in Fig. 5-9. You'll probably have a loss of color. When both video heads are badly clogged, the entire picture is completely snowy.

There are a few VCRs that mute the video signal and cause the TV screen to go completely black. In newer models the screen will be completely blue. When viewing a broadcasted channel from the VCR, the picture is okay, but when inserting a video tape and pushing Play, the TV screen turns completely black or blue. If, however, the audio portion of the video tape sounds okay, usually the cause is a clogged video head.

A four-headed machine

Let's say that one of the four heads is clogged and you inserted a two-hour video tape and pushed Play. You find the image on the TV screen is okay, but when you insert a four- or six-hour video tape, the image is distorted or snowy. On the other hand, the two-hour tape could

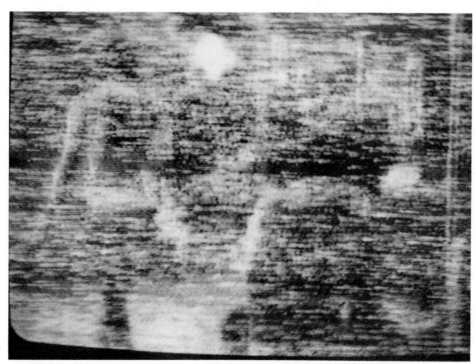

5-9 A partially distorted picture.

look snowy or distorted, but the six-hour tape looks okay. What I'm saying is this: two of the video heads affect the six to four hour video tapes, while the other two heads affect the two hour video tapes. To correct this problem, you'll need to clean all four heads, as previously explained. After cleaning the video heads, if you are still receiving a distorted picture, clean it again. I've been able to save many hours of frustration and much money by recleaning the video heads.

Review

☐ Remember all video drums look the same.

☐ The drum has two parts:
  - The upper video drum rotates and holds the video heads.
  - The lower video drum is stationary and supports the upper drum.

☐ Never remove or disassemble the video drum.

☐ You need to clean the upper drum and the tips of the video heads.

☐ Saturate a clean chamois stick with cleaning solution.

☐ Move the chamois stick up and off the video drum in one motion at a slight angle, not up and down.

This chapter has given you a lot of information to absorb. You should be feeling comfortable with the information and your capabilities. The next chapter is on saving a video head. Cleaning the video heads, as well as the other routine cleaning procedures, comes down to proper maintenance.

6
Pinpointing and correcting video head problems

IF YOU PROPERLY CLEAN THE VIDEO HEADS AND STILL HAVE NO PICTURE, BUT only snow on the TV screen, the next step is to determine if the VCR has a bad video head or a circuit problem.

Detecting a circuit problem

Find the TV-Video switch on the face of the VCR and place it to the video position. Then tune in a broadcast channel using the tuner on the VCR. If you can properly fine-tune a channel, 90% of the unit is working correctly. You've eliminated all circuit problems, except for the video head amplifier and the video heads. The next step is to determine if the problem is in the video head amplifier.

Finding the video leads

Locate the video leads to check the video amplifier. Usually one to four large round leads are coming from the bottom rear portion of the lower video drum. The shielded cable houses the video leads. Leads are usually gray or white in color, but are red or yellow in a few models.

Older models have two large round leads soldered to a small circuit board, mounted right behind and on the lower video drum. (See Fig. 6-1. I've removed the video drum to give you a better view.) In a few models, the circuit board is covered with a small shield plate held on with one screw. You'll need to remove this plate.

A few years after those models were made, the manufacturer's started using a video plug instead of soldering the video leads to a small circuit board. In these models, you'll find two large round leads protruding from a white plug mounted at the rear of the lower video drum. In some models, the white plug is at the right rear portion of the video drum, as shown in Fig. 6-2. In other models, the plug is horizontally placed right behind the video drum.

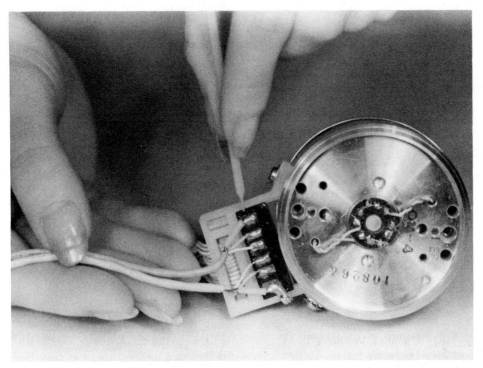

6-1 Locating the video leads and connections.

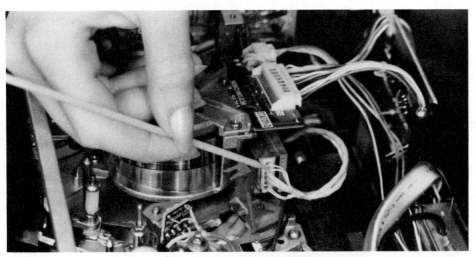

6-2 Locating the video plug.

In stereo hi-fi models, you might find three large round leads coming from a white video plug at the rear of the lower video drum, as shown in Fig. 6-3. Follow the leads coming from the video plug. Two of the leads pair up and go to a small covered metal box on a circuit board, usually located near the top of the unit. These two leads are the video leads. The third

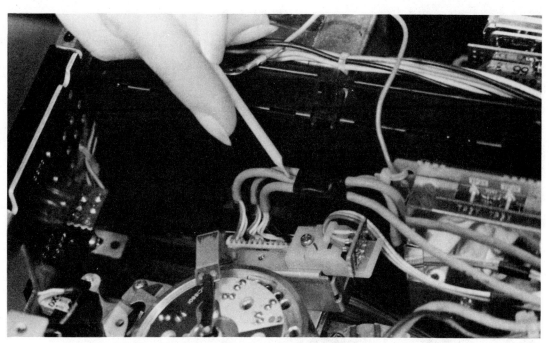

6-3 Locating the three shielded video leads in stereo hi-fi models.

lead runs to a different circuit board, usually near the bottom of the unit. This lead is for the stereo hi-fi, or audio section.

In other stereo hi-fi models, you'll find four large round leads coming from the video plug (Fig. 6-4). Follow these leads. They pair off and go into two different sections of the unit. One pair of leads goes to a covered metal box on a circuit board, usually near the top of the unit. These leads are the video leads. The other pair of leads usually goes to a circuit board, near the bottom of the unit. These leads are the audio leads.

A few models have a rotary cap on top of the upper video drum. Two white plugs connected to a circuit board are attached to the back of the rotary cap. Do not confuse these two plugs with the video plug. The plug to the right of the lower video drum (Fig. 6-5) is the video plug and has only two large round leads coming off it. This is the plug you're looking for.

One more system is used in the most current models. Instead of using large round leads, they now use a ribbon. A ribbon is a group of wires inside one insulation and is flat in appearance. The ribbon protrudes out from behind the lower video drum. At the opposite end of this ribbon are exposed bare contacts, where the ribbon is folded back, as shown in Fig. 6-6. These are the video leads, or connections. In other models, the ribbon goes into a covered metal box right behind the video drum. This ribbon is connected to a plug on the top or front of the box.

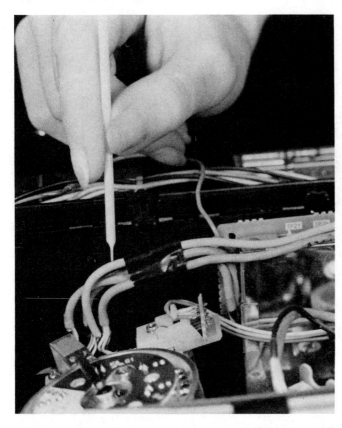

6-4 Locating the four shielded video leads in stereo hi-fi models.

Checking the video head amplifier

No matter what model you have, find the best previous illustration for locating the video leads. Your VCR could be the type with large round leads soldered to a circuit board, with leads coming from the white video plug, or with the ribbon.

Now that you've located the video leads in the VCR, start with units using a video plug, whether or not you have a stereo hi-fi. At the end of each large round lead, just before it goes into the video plug, are three small wires connected to each large lead. You'll find either a blue or a yellow ground wire. You'll also see a red wire and a white wire. These two are the video wires. All three wires are attached to small pins inside the video plug. The pins connected to the red and white video wires are the ones you are concerned with.

Make sure the TV-Video switch is in the video position. Insert a video tape and press Play. You should see snow on the TV screen, as shown in Fig. 6-7. Find the point the red and the white wires enter the white plug and connect to the pins. Press your finger against the metal shaft of a small flathead screwdriver. Touch the corner of the tip to each metal pin one at a time, starting at one end of the plug and working across it. Touch the red and the white wires, as shown in Fig. 6-8. Touch all four video pins. If you're not sure which pins are the video pins, touch all the pins on the video plug. All pins on the video plug carry no voltage and will not shock you.

6-5 The rotary cap plugs on top, and the video plug on the side.

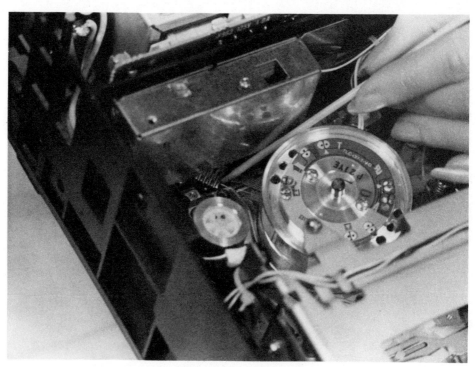

6-6 Bare contacts on the ribbon.

6-7   A snowy screen.

6-8   Touching the pins in a video plug.

6-9 A black band appearing on the screen.

Keep an eye on the TV screen. Each time you touch a red or black pin directly connected to the input of the video amplifier, the snow on the TV screen changes to many black bars or a wide black band, as shown in Fig. 6-9.

In other units, you might have white bars instead of black bars, or the snow changes from a light normal snow (Fig. 6-7) to a heavy darker snow (Fig. 6-10). Each time you touch a video amplifier pin, a definite change occurs on the TV monitor. When the screwdriver is on a video pin and you touch the screwdriver, you become an antenna, causing lines to appear on the TV screen.

Remember that only two of the pins on the video plug change the image on the screen. All the other pins on the plug have no effect on the image. Also, be careful not to short any two pins together.

If you're working on an older unit, the red and the white wires are soldered to a small circuit board behind the lower video drum. Touch a screwdriver to the circuit board where the red and the white wires are connected, one connection at a time (Fig. 6-11). With this model, both the red and the white wires from each video lead change the image on the TV monitor.

If you have a ribbon type unit, touch each exposed connection, one at a time. Only two connections on the ribbon change the image on the TV monitor. Figure 6-6 shows the location of these contacts. None of the connections carry voltage, so they will not shock you. If you have a ribbon that goes into a covered metal box, touch each pin where the ribbon goes into the plug.

If you found at least one wire from each video lead that changed the image on the TV

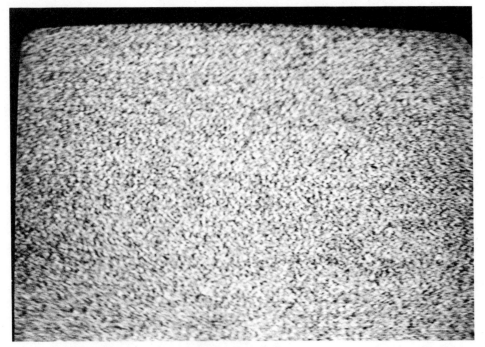

6-10 Heavy snow.

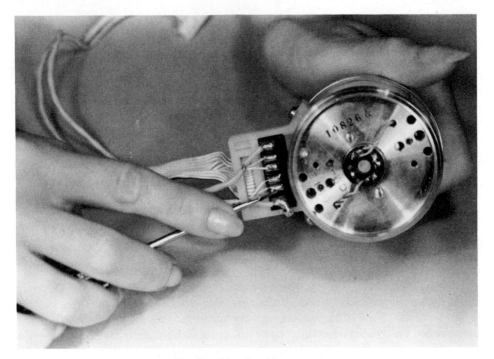

6-11 Touching the video contacts.

monitor, the video amplifier is functioning. You've also eliminated all circuit problems in the VCR. You now know the problem is in the video heads.

Take a real close look at each video head. If you have a magnifying glass, use it. Check the tip of each video head protruding through the slot. See if it might be cracked, chipped, or broken off. Compare all the video heads on the upper drum to see if they appear exactly the same. If they do, the problem is a badly clogged video head or heads. Taking the unit into a service center at this time could prove to be quite costly. The service personnel wouldn't go any further with this problem, other than to replace the whole video drum or heads. There is another option.

Saving the video head

Follow the normal procedure of cleaning the video heads. Use a chamois stick to clean it at least five or six different times. If you still have no picture after cleaning the head with the chamois stick, it's time to try something new.

You need a fiberglass brush. The fibers on the brush penetrate the viewport and pull out the dirt. Use a chamois stick to clean the outside of the video drum and the tip of each video head, or viewport.

Before using the fiberglass brush, make sure you clean it by dipping it into the cleaning solution. Take a paper towel and wipe off any dirt. The paper towel also removes any loose fibers left on the glass brush. Clean the brush several times, as shown in Fig. 6-12.

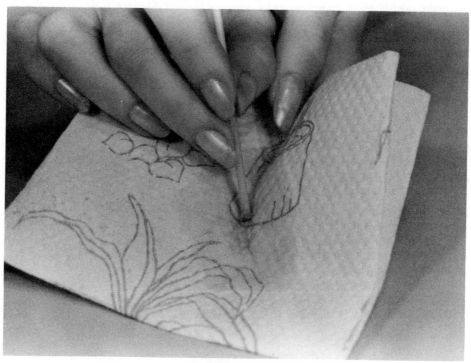

6-12   Cleaning the glass brush.

6-13   Placing the glass brush in the slot.

Face the front of the VCR towards you. Go to the upper video drum and rotate it until one of the video heads is facing towards the front right side of the video drum. Hold the brush at a 45-degree angle and place the brush directly behind the viewport and into the slot (Fig. 6-13). Move the brush forward across the viewport, staying in the slot. Then stop and pull the brush away from the slot. Repeat this again staying in the slot, carefully moving the brush in only one direction across the viewport. Repeat this procedure 20 or 30 times.

Rotate the upper drum and clean the next video head. Be sure to clean all the video heads on the upper drum. Do not let the brush rub on the drum itself or you'll scratch it. Do not work the brush in an up and down motion or you will break the video head. When you've finished, take a saturated chamois stick and clean the upper video drum to remove any loose fibers left on the drum. (Refer to the section in chapter 5 on "Cleaning the video heads.")

You are now ready to insert a video cassette. Press Play to see if you have a picture. If not, repeat the cleaning procedure several more times. If this procedure fails to work, there is yet another method; however, you must be careful. Instead of sweeping the video head as previously stated, use a scrubbing motion, back and forth only. I've found this to definitely work as long as the video head has not been damaged. Don't give up, eventually you'll get a partial picture, as shown in Fig. 6-14. This is due to particles still remaining in the viewport. When you reach this point, do not clean it again. Let your VCR play for an hour or so. The video head will clean itself out.

Cautions regarding the video drum

There are three main concerns when you're working on or around the video drum.

Never touch the sides of the drum, you can leave oil smudges on the drum that can clog the video heads.

6-14   A partial picture.

When working around the video drum while the unit is running, never touch the drum with anything.

When you go into the Stop mode, be sure to let the upper drum stop spinning completely before you proceed to work on the VCR.

Review

☐ First try cleaning the video heads five or six times using a chamois stick.

☐ You need to eliminate the video amplifier. Place your finger on the shaft of a small screwdriver and touch the tip of the screwdriver as follows:
- To the pins on a white video plug.
- To the connection on the circuit board right behind the lower video drum.
- Where a ribbon connects to pins on a video plug.
- Where a ribbon is folded back and solder makes connections.

☐ After finding a video lead, look at the TV monitor to see if the image on the screen changes. If so, the video amplifier is functioning properly.

☐ Clean the video heads by carefully using the glass brush.

☐ After using the glass brush, don't forget to reclean the video drum with a saturated chamois stick.

☐ After you finally receive a partial picture on the TV screen, do not clean the video heads again. You could make them worse. Let the VCR play and let it run until it cleans itself out.

☐ Don't give up! You can save the video heads.

7
Servicing and lubricating VCRs

IF A VCR IS A FEW YEARS OLD OR HAS A LOT OF HOURS ON IT, IT'S TIME TO lubricate the unit, especially if degreaser has been used to clean the cassette carriage. Degreaser removes all the oil in the carriage and leaves the lubricating points dry. Use a high grade of household oil to lubricate. Do not use a spray lubricant; it leaves an oily film on the belts and pulleys that causes slippage. Do not get any oil on the video drum or on any part that comes into direct contact with the video tape.

Finding the lubricating points

Plug in the unit, insert a video tape, and push Play. Keep an eye on the transport. As the video tape loads onto its tape path, you'll be able to see how all the parts operate. Many parts are interconnected, usually by a hinge. An E- or C-ring is at each hinge. Lubricate them, as shown in Fig. 7-1. Stop and start the unit two or three times to find all the lubricating points. Then remove the video cassette. Put a small drop of household oil onto each hinge where the parts interconnect. Lubricate all hinges. Do not lubricate the video drum or get any oil on the pulleys or belts.

To find the lubricating points under the cassette carriage, the cassette holder has to be in the up position. Push Eject. Now you can look under the cassette holder to see all the other moving parts. Using your finger, wiggle each part back and forth one at a time. This will show you where the parts are connected or hinged. Place a small drop of household oil onto each hinge. Remember that each unit has a different combination of parts inside, but the types of hinges remain the same.

Older models have hinged mechanical function buttons that need to be lubricated. Push down each button one at a time to expose the hinges. Lubricate as indicated in Fig. 7-2.

Lubricate the shaft on the idler wheel, the pinch roller, and the tension rollers (Fig. 7-3). The shaft is a metal pin that protrudes through the middle of the wheel or roller. On top

7-1 A lubricating point.

of each shaft is a C- or E-ring. After lubricating each shaft, spin each roller. The rollers should spin freely for a second or two. If a roller is hard to turn and won't spin freely, you'll need to pull the roller off its shaft. The idler wheel should turn easily but not spin like the rollers.

To remove a wheel or roller, remove the C- or E-ring connected to the top of the shaft. Some idler wheels don't have a C- or E-ring. If this is the case, proceed to the section in chapter 29 on "Removing idler arms and clutch assemblies." After removing the ring, pull the wheel or roller straight up and off. Be careful; a small washer is on top and/or under the wheel or roller. For removing a pinch roller, proceed to the appropriate section in this chapter.

Removing E-rings

To remove an E-ring, place a small flathead screwdriver in the space between the ring and the shaft (Fig. 7-4). Place your finger behind the ring, so when the ring comes off the shaft, it won't fly across the room. Twist the screwdriver and the ring will pop right off. After removing the ring, pull the roller or wheel straight up and off the shaft. You might find a small washer on the top and bottom of each roller or wheel. Be sure not to lose these washers and

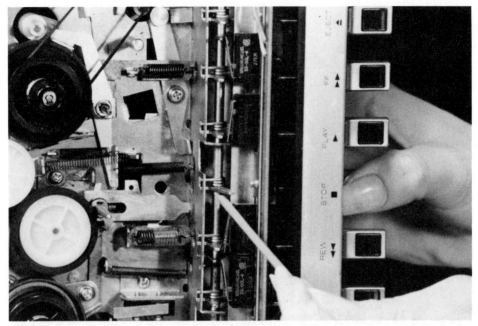

7-2 A mechanical function button hinge.

7-3 The idler wheel (bottom), the pinch roller (center), and the tension rollers (top).

7-4 Removing an E-ring off the shaft.

to put them back on the shaft when reassembling. To put the E-ring back onto the shaft, place the open part of the E-ring into the notch in the shaft. Take a pair of long nose pliers and snap the E-ring onto the shaft, as shown in Fig. 7-5.

Removing C-rings

Removing C-rings is more difficult because there is tension on the ring. You have two options for removal: purchase a C-ring puller from a local electronics supply center, or use a flathead screwdriver to pry up the ring, as shown in Fig. 7-6. To replace the C-ring, place the ring on top of the shaft, holding one end down with your finger. Take a flathead screwdriver and pry one side of the split in the ring down over the shaft (Fig. 7-7). Take the screwdriver, place it over the other side of the split in the ring, and push as before, placing the whole ring onto the shaft. Place the screwdriver onto both sides of the split end, along with a finger, and push the C-ring down the shaft until the wheel or roller is secured.

Removing a pinch roller

To remove the pinch roller, place a Phillips head screwdriver into the top hole and remove the screw. The location of the screw is shown in Fig. 7-8. Pull the pinch roller straight up and off the shaft. Be aware that some pinch rollers have a white cap on top. To remove the cap, place a small flathead screwdriver between the cap and the top of the pinch roller. Twist the screwdriver and the cap pops right off (Fig. 7-9). Remove the screw. In a few models, all that holds the pinch roller on is the white cap.

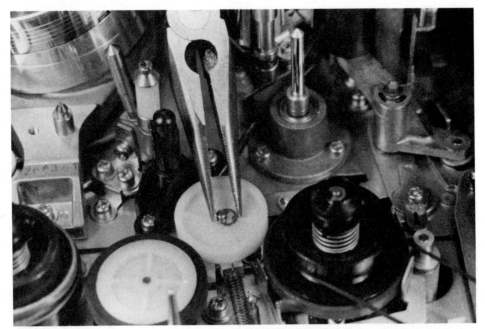

7-5 Remounting an E-ring onto the shaft.

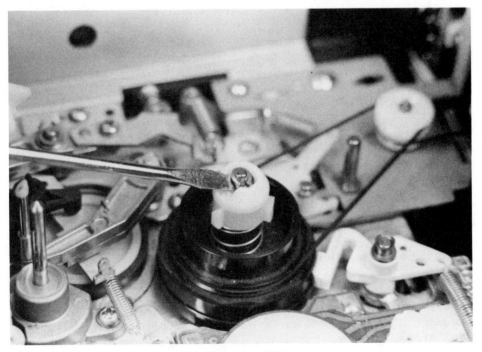

7-6 Removing a C-ring.

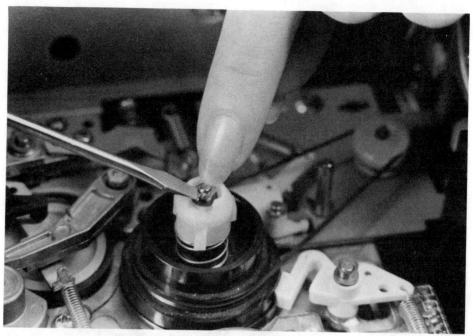

7-7 Remounting a C-ring.

7-8 Locating the pinch roller mounting screw.

7-9   Removing the white cap on top of the pinch roller.

Wheels that do not turn freely

For demonstration purposes, I'm using the pinch roller as a part that doesn't spin freely. After removing the pinch roller, you'll find a buildup of black residue on the pinch roller shaft. Take a can of degreaser and spray off the residue onto a papertowel, as shown in Fig. 7-10. Wipe off any excess residue with a clean papertowel. Clean out the inside of the pinch roller. Take the degreaser and spray down inside the pinch roller. Take a glass brush and spray some degreaser onto it. Take the brush and scrub the inside of the pinch roller until it's free of all residue, as shown in Fig. 7-11. Spray the inside of the roller again to remove all fibers left by the brush. Place a drop or two of oil onto the pinch roller shaft, then place the pinch roller back onto the shaft. Replace the screw and cap. This procedure is used for all rollers and wheels that don't turn freely.

The roller guide tracks

There are parts in the unit that need to be greased instead of oiled. Underneath the roller guides, you'll find tracks, as shown in Fig. 7-12. Relubricate these tracks if you used degreaser on them, or if the grease has been rubbed off by the roller guides. If the old grease becomes black and hard, you'll need to remove it. Remove the black grease with a paper towel, or spray it off with degreaser. Any caked on grease spots left on the tracks can be removed using a flathead screwdriver wrapped in a paper towel (Fig. 7-13).

On these tracks, use tuner grease or phono lube. To reapply the grease onto the tracks,

7-10 Removing residue from the pinch roller shaft.

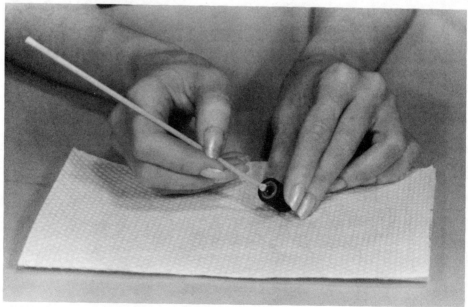

7-11 Using a glass brush to clean inside the pinch roller.

7-12   Location of the roller guides tracks.

7-13   Using a papertowel to remove old grease.

7-14 Applying new grease to the tracks.

dip a small flathead screwdriver into the grease, and apply it to the tracks, as shown in Fig. 7-14.

After applying the grease insert a video cassette, push Play and then Stop. Repeat this several times, so the roller guides distribute the grease evenly onto the tracks. Remove any excess grease with a paper towel. You don't lubricate some moving parts, including the rollers on the roller guides and the video drum. Also, never over lubricate. A little bit goes a long way.

Review

☐ To find all lubricating points:
  • Turn the power on, insert a video tape, push Play, and make note of all the parts that move, especially the hinges.
  • Move the cassette holder up, wiggle each piece under the cassette holder with your finger, and look for all the hinges.
  • Spin all rollers and wheels to check for freedom of movement.

☐ Put a drop of household oil onto each hinge, roller, and wheel shaft.

☐ Do not over lubricate. Just one drop will do.

☐ Do not lubricate the video drum.

☐ Regrease the tracks under the roller guides, if needed.

8
Children and VCRs

SMALL CHILDREN CONSIDER A VCR THEIR PIGGY BANK. THEY LIKE TO LISTEN to coins drop through the air vents or the cassette door. I've removed enough coins over the last five years to fill a large cereal bowl.

A favorite pastime for children is to stuff small toys and crayons through the cassette door. I've found items, such as pen caps and crayons (Fig . 8-1), that've jammed the unit. They have fallen into one of the roller guide tracks and stopped the unit from loading the video tape. A small toy might float around inside the unit and eventually find its way into the teeth of some gear and jam it, as shown in Fig. 8-2. A dragon fly is jamming the teeth of a gear. If a VCR is making strange noises or sounds like something is straining, remove the main top cover and go treasure hunting. You'll be amazed at what you can find in a child's VCR piggy bank.

Coins conduct electricity, and when they find their way to a circuit board they short it out. If a section of the VCR goes dead, check around on all circuit boards for coins or any other foreign objects made of metal and remove them. Then check for a blown fuse. In some cases the foreign object might have shorted out a B-plus line only, which causes the fuse to turn black inside. A B-plus line is a source that feeds electricity to select sections in a VCR. Replace this fuse. (For replacing fuses, refer to chapter 12, "Replacing Fuses.")

One day while trying to view a movie, a horizontal belt of lines appears on your TV screen. You find the tracking adjustment has no effect. Remove the top cover and check for a straw or piece of paper caught between one of the roller guides and the V-mount, as shown in Fig. 8-3. Just a small piece of paper caught behind one of the roller guides can throw the guide out of alignment.

If you aren't sure if a child has made a deposit in the VCR piggy bank, pick up the unit and shake it back and forth. Listen for a rattling sound coming from the inside of the VCR. If you can't find the foreign object inside the unit but you still can hear a rattling sound when shaking it, turn the unit upside down and shake it until the foreign object falls out.

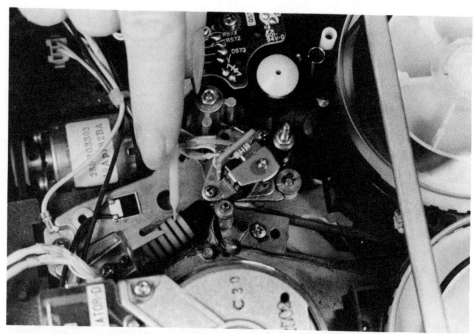

8-1   A foreign object jamming the roller guides.

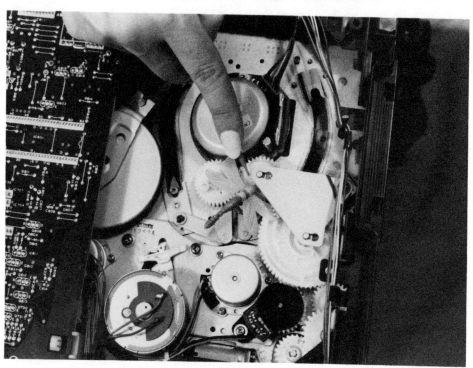

8-2   A foreign object caught in the gearing.

8-3 A foreign object caught in the V-mount.

9

Jammed video cassettes

A VIDEO CASSETTE IS THE CARTRIDGE WHICH HOLDS THE VIDEO TAPE. THE cassette holder is what you insert the tape into. The cassette holder pulls the tape into the machine in frontload VCRs. The cassette carriage is the complete assembly that cradles the cassette holder and is attached to the transport. These are the main parts covered in this chapter.

Reasons why the cassette holder jams

The cassette holder in your VCR can become jammed for several reasons. These reasons and the means to remove the jammed cassette are discussed in this section. Read each of the reasons and use the one that fits your situation.

First reason

I've come across many instances in which a video cassette has been left in the unit after viewing. In a day or two, you insert another video cassette. It goes in, but only flush with the front panel. The VCR won't pull the video cassette into the unit. You push Eject to remove the video cassette, but the cassette won't eject. You try to pull the video cassette back out through the cassette door using your fingers or some other object only to find the video cassette is jammed in tight and can't be removed. Does this phenomenal situation sound familiar? What's a person to do?

Open the unit by removing the main top cover. Look at the cassette holder and you'll see the jammed video cassette. Place two fingers behind the video cassette (Fig. 9-1) and gently push the video cassette out through the front opening.

In some VCRs, the video cassette can hang up on top the cassette holder. Pull up on the top bracket of the cassette holder and simultaneously push the video cassette out, as shown in Fig. 9-2.

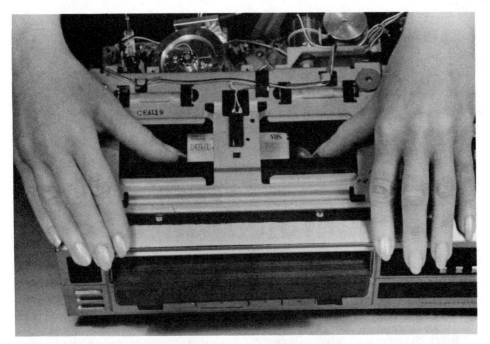

9-1 Pushing out a jammed video cassette.

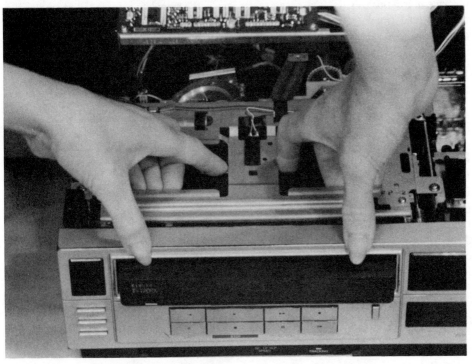

9-2 Pulling up on the bracket to remove the video cassette.

9-3 The video cassette indicator light.

Now that you've gotten the video cassette out, why did it get jammed? Push the Eject button and see what happens. Another video cassette comes out. Hardly anyone ever checks to see if a video cassette was left inside the unit. To avoid this happening, check the front of the VCR and look at the video cassette tape indicator. The tape indicator lights up when a video cassette is inserted (Fig. 9-3).

Second reason

You might push Eject and the video cassette starts to come up, but suddenly it stops and goes back down, or the video cassette almost comes completely out, but stops and goes back in. On the other hand, you might try to insert a video cassette and it comes right back out. What you might find is a foreign object caught in the teeth of the gears or the worm gear that makes the cassette carriage move. Look closely at these gears. All it takes is a very small piece of plastic, wood, or metal caught between two teeth or in the worm gear (Fig. 9-4). The reason the gears jam is because the gears usually have grease on them, and when a very small foreign object falls onto the gear, it'll stick to the grease. When the gears start moving and come across the foreign object, the object keeps the gears from turning. The objects can be as small as a pin head and still jam the gears.

To check the teeth for a foreign object, first turn off the unit. Remove the cassette carriage. (Refer to chapter 16, "Removing cassette carriages in topload VCRs," or chapter 18, "Removing cassette carriages in frontload VCRs." Next look to the back on the right side of the cassette carriage for a drive pulley with a drive belt on it. Place your finger on the drive pulley and turn it. At this point, the pulley should turn in a clockwise direction only. As you turn the pulley, all the gears will turn, as shown in Fig. 9-5. When the gears no longer turn, look at where the teeth come together on each gear on both sides of the cassette holder. Make a mental note of this position. This location is where you'll find the foreign

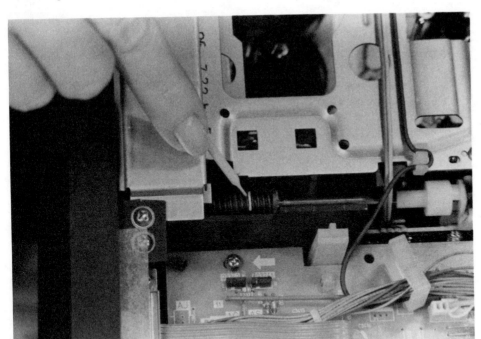

9-4   A foreign object caught in the gearing.

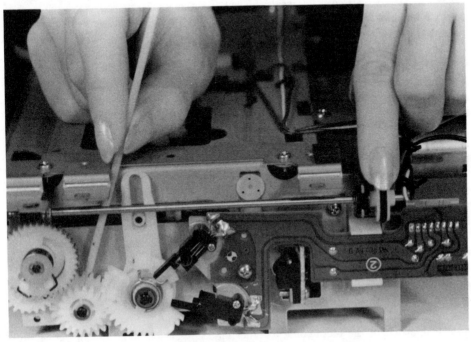

9-5   Turning the pulley to make the gears rotate.

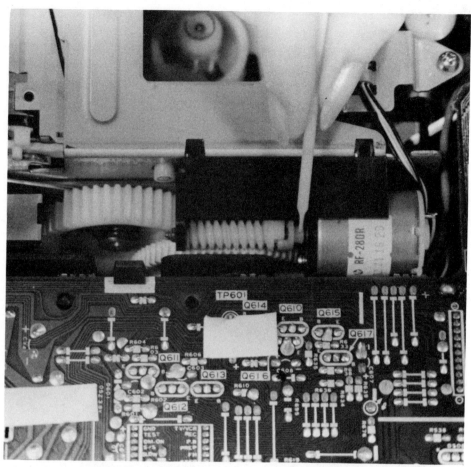

9-6 The worm gear attached to a housing loading motor.

object. Turn the pulley counterclockwise to back up the gears and expose the object for removal.

In some models, you won't find a drive pulley in the right rear corner of the cassette holder. Instead, you'll find a worm gear attached directly to the motor, located on the right of the cassette holder (Fig. 9-6). Use your finger or the tip of a flathead screwdriver to rotate the base of the worm gear. All the gears will move, allowing you to locate the foreign object.

Third reason

When you push Eject, you hear the sound the unit makes when ejecting a tape, but no video cassette comes out of the unit and the sound stops. The problem is a broken drive belt. This belt is on a pulley, near the rear of the cassette carriage. In this case, just replace the broken belt. Proceed to the first four sections of chapter 26 for detailed instructions.

Fourth reason

You insert a video cassette, you hear a squealing sound, and the tape comes back out. This condition is caused by a bad drive belt, which is connected to the same pulley, as shown in Fig. 9-5. Replace the belt.

Fifth reason

If you push Eject and receive absolutely no response, this condition can be caused by a bad eject motor. Proceed to the sections in chapter 27 on "Checking motors for dead spots" and "Locating frozen motors."

Sixth reason

You insert a video cassette, but it only goes in part way and then jams. All the mode buttons are nonfunctional, so you unplug and plug in the unit to reset the microcomputer. The video cassette will automatically eject. (In other models, you'll have to manually turn the pulley or the worm gear to eject the video cassette.) After retrieving the video cassette, turn the power off. Insert the video cassette by placing a finger on each end of the cassette and

9-7   Aligning the dots on a gear.

pushing it in. The cassette should insert evenly. If one side goes in easily and the other locks, then there's a broken gear on the carriage. Refer to chapter 18, "Removing cassette carriages in frontload VCRs." After removing the carriage, find the broken gear. To remove a gear, take off the C-or E-ring and pull off the gear. To replace a gear, the cassette holder has to be in the up position. Each gear has an arrow or a dot, as shown in Fig. 9-7. To replace the gear, align the arrows or dots toward each other, and then replace the C-or E-ring.

Seventh reason

You're watching a rented movie and suddenly the VCR starts speeding up, then the unit stops playing. You try to push the play button again, but the unit still won't work. You decide the unit is broken, and you want to take the rented movie out. You push the Eject button, but the video cassette doesn't eject. Before you take the unit into a local service center for repairs, let me explain a common problem that occurs with VCRs.

Take off the main top cover. You'll probably find a ball of video tape wrapped around the capstan shaft, as shown in Fig. 9-8. The reason this happens is that rental tapes take a lot of abuse from previous renters. The tapes can become dirty or sticky from spilled beverages, causing the tape to stick to the capstan shaft and roll up in a ball.

When you find a ball of video tape, it usually is wrapped around the capstan shaft until it's wedged tightly between the capstan shaft and the front cover of the video cassette. (See Fig. 9-9. I've removed the top cover of the video cassette to give you a better view.)

To remove the video cassette, remove the ball of video tape wrapped around the capstan shaft. Take a flathead screwdriver and place it directly under the ball of video tape. Now

9-8   A ball of video tape wrapped around the capstan shaft.

9-9 A ball of video tape wedged against the video cassette.

twist the tip of the screwdriver back and forth between the base of the capstan shaft and the bottom of the ball of video tape. This works the ball of video tape, up the capstan shaft. Continue to work the ball of video tape up the shaft for at least ¼ inch. If you have a topload unit, push Eject. The cassette carriage should pop up ½ inch or so. Carefully find a solid bracket to lay the shaft of the screwdriver on and pry the ball of video tape up and off the capstan shaft, as shown in Fig. 9-10. As the ball of video tape comes off the capstan shaft, the cassette carriage is able to pop up all the way. You now can remove the video cassette.

9-10 Prying the ball of video tape up the capstan shaft.

If you have a frontload unit, find a solid bracket to lay the shaft of the screwdriver on to pry the ball of video tape up and off the capstan shaft. In this model, push the Eject button and pry the ball of video tape up the shaft at the same time. Frontload units are motor driven, and the motor must be running to pry the video tape off. You might need to repeat this procedure several times before the cassette holder will eject all the way. Be careful to place the shaft of the screwdriver on a solid bracket when prying up on the ball of video tape. *Do not place the screwdriver shaft on any part that can be bent or broken.*

Usually the ball of video tape comes off the capstan shaft fairly easily. Sometimes, however, the ball of video tape is wrapped around the capstan shaft so tightly you can't work it up the shaft with a screwdriver. In this case, take a single-edged razor blade and cut the ball of video tape off the capstan shaft. Slice it in the same place until you cut clear down to the metal shaft, as shown in Fig. 9-11. Push Eject, and the video cassette will eject. Sometimes the cassette carriage won't eject because of the remaining tape between the capstan shaft, and the front of the video cassette. Simply place a screwdriver under the video cassette near the capstan shaft and pry up (Fig. 9-12). Be sure to push Eject at the same time you are prying up on the cassette. The video cassette should eject fairly easy, with the aid of a screwdriver. After removing the video cassette, finish unwrapping the rest of the video tape on the capstan shaft. Remove all the cut pieces of video tape from the unit. Then refer back to the section in chapter 4 on "Capstan shaft" under the subhead "Cleaning the various components" and to chapter 11, "Repairing a video tape."

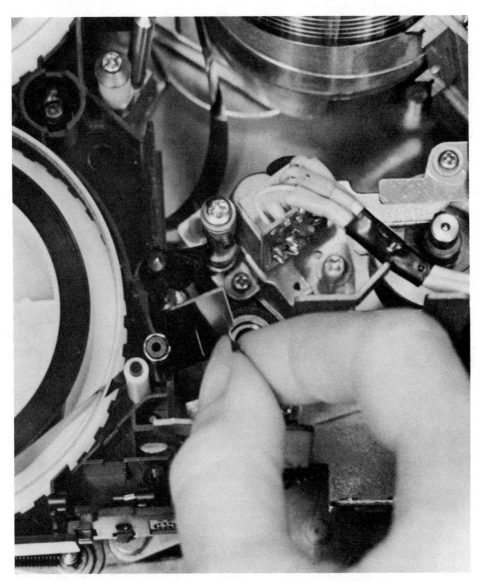

9-11 Slicing the ball of video tape with a razor blade.

Video cassette won't insert in frontload VCRs

You've inserted a video cassette in crooked, and the cassette holder started to pull the tape in. Because it was inserted crooked, the video cassette didn't move with the cassette holder. Eject won't eject the tape. The cause of this problem is that the locking latches have locked in the cassette carriage. To correct this problem, unplug the unit, so that the microcomputer resets. Plug the unit in, and the tape will automatically eject.

After you've turned the power back on, the cassette holder will move in about an inch,

9-12   Prying up on the bottom of the video cassette.

so you'll be unable to insert a video cassette. At the same time the cassette holder moves in, the cassette light indicator comes on. The cassette holder will either stay in this position or it'll pop back and forth. If you are unable to insert a video cassette, unplug the unit and plug it back in to reset the microcomputer. This procedure returns the cassette holder to its loading position.

If you insert a video cassette and it automatically ejects after two to five seconds, or if you press Eject but the tape is pulled back into the unit before you can retrieve it, bad tape sensors are the cause of the problem. Proceed to chapter 18, "Removing cassette carriages in frontload VCRs." Then read the section in chapter 15 on "Replacing tape sensors" under the subhead "Frontload VCRs." Replace both of the tape sensors.

In one frontload model, you'll find two wheels on top of the cassette carriage that look like two pinch rollers. This unit is the only chain driven model on the market. A chain drives the two rubber wheels. When you insert a video cassette, these wheels turn, pulling the cassette in. When the rubber on these wheels is hard or shiny or has cracks, the wheels will slip and not pull in the cassette. To correct this problem, replace the wheels or proceed to the section in chapter 28 on "cleaning rubber tires."

A jammed video cassette in a topload VCR

You have a jammed video cassette inside the VCR. You've removed the main top cover screws; however, the problem is getting the top cover off to get inside. The jammed video cassette inside the unit is holding down the cassette lid. The cassette lid has a lip around it, which is in turn holding down the main top cover.

9-13 Locating the cassette lid screws.

Removing a cassette lid

After you've removed all the screws from the main top cover, look for two screws on the top of the cassette lid, as shown in Fig. 9-13. Remove the screws, place your fingertips at the rear of the cassette lid, and push the lid forward, so the lid moves towards the front of the unit. In some models, you'll have to push hard. Push the lid as far as you can, while pushing down on the back of the cassette lid. The front latches will release, and the front of the cassette lid will pop up, as shown in Fig. 9-14. Remove the lid, and the top cover comes right off.

In some models, there are no screws on top of the cassette lid. The lid has to be in the up position to remove the main top cover. To get the lid to come up, try pushing Eject. Sometimes the cassette lid will pop up enough to grab hold of it. If not, take either a wide putty knife or a kitchen spatula and place it under the cassette lid (Fig. 9-15). Carefully pry the lid up each time you push on the Eject button. After you have gotten the cassette lid to come up a ¼ inch or so, grab hold of it, as shown in Fig. 9-16. Now pull up on the cassette lid using a jerking motion until the lid comes up enough to remove the screws on each side of the cassette lid (Fig. 9-17). In this case, only one screw is on each side. After removing the screws, push forward on the lid to release the front latches. Pick the lid straight up and off.

Other cassette lids have two screws on each side. Just remove the four screws and pick the lid straight up and off. On all other models, the main top cover comes off without having to remove the cassette lid first. After removing the top cover, follow the procedure on how to remove a jammed video cassette.

9-14 Pushing on the cassette lid to release the latches.

9-15 Prying open the cassette lid.

9-16 Pulling the cassette holder up.

9-17 Removing the screws on the sides of the cassette lid.

Manually removing video cassettes

There's a video cassette in the unit, and Eject won't eject the video cassette. Remove the main top cover and locate the pulley at the right rear corner of the cassette carriage. This pulley (Fig. 9-5) has a drive belt attached. Turn the pulley clockwise, making all the gears on the carriage rotate. Continue this process until the video cassette ejects. You might find a worm gear instead of a pulley located on the right side of the carriage. Rotate the base of the worm gear clockwise until the video cassette ejects. This process ejects the cassette, unless the unit is stuck in play.

To remove a video cassette when the unit is stuck in the play mode, refer to chapter 25, "Getting into the undercarriage." Then place the unit on its side and locate the loading motor. (Refer to the section in chapter 27 on "Removing loading motors.") Rotate the pulley or worm gear on this motor clockwise until the roller guides on top of the transport have completely retracked. Locate the largest wheel in the undercarriage that is connected to the capstan shaft. Turn the wheel in either direction until it pulls the video tape back into the cassette. Then, follow the procedure in the last paragraph.

Review

☐ On frontload units, remember to check that the tape indicator lamp is off before inserting a video tape.

☐ Always insert a video cassette in straight, not crooked.

☐ On frontload units, when the video cassette doesn't eject, check for a broken or bad drive belt, then look for foreign objects caught in the teeth of the gears.

☐ You're viewing a video, and suddenly the VCR stops and you cannot eject the video cassette. Check the capstan shaft for a ball of video tape wrapped around the shaft.

☐ When removing the video tape from the capstan shaft, use a screwdriver and be careful not to bend or break any other parts.

☐ In most topload units, you need to remove the cassette lid to remove the main top cover.

10

Video-tape-eating VCRs

IT'S SO FRUSTRATING TO RENT A MOVIE, POP SOME POPCORN, AND INSERT the movie for viewing, only to have the VCR suddenly eat the video tape. Sound familiar? This chapter covers where and why this annoying problem happens and how to remove the damaged tape from your VCR.

Two locations at which a video tape can be eaten

There are two general locations where your VCR is most likely to try to eat a video tape. The first location is between the pinch roller and the capstan shaft. The other location is on the take-up tape guide. This section will show you how to remove the tape if it is caught in either of these two locations.

First location

The video tape goes between the pinch roller and the capstan shaft. (See Fig. 10-1. I've taken off the top cover and door of the video cassette to give you a better view. You don't need to remove these parts.) If the pinch roller and the capstan shaft aren't aligned exactly parallel to each other, the video tape will be eaten.

Second location

Another location that a video tape can be eaten is on the take-up tape guide. This tape guide is between the A/C head and the capstan shaft. Normally, the video tape flows smoothly across the tape guide. In some cases, however, there's pressure on the video tape, pushing it against the bottom edge of the tape guide. This pressure causes a crease or a fold in the video tape. The fold travels between the capstan shaft and pinch roller, causing the bottom edge of the video tape to be eaten. The bottom edge of the take-up guide causes the fold, as

10-1 Video tape between the pinch roller and capstan shaft.

shown in Fig. 10-2. The A/C head is not properly aligned with the video tape. This misalignment puts pressure on the tape in a downward motion. Do not touch the adjustment nut on top of the take-up tape guide. Push Play and locate the allen nut or adjustment screw located directly behind the A/C head, as shown in Fig. 10-3. This tilt adjustment screw causes the A/C head to change its contact angle with the video tape by either tilting the A/C head forward or backwards.

The tilt adjustment is either a Phillips-head screw or a 1.5mm allen nut. Place the proper tool in the rear nut or screw and begin turning it counterclockwise. As you slowly turn the tilt adjustment, the A/C head starts tilting backwards and changing its contact angle with the video tape. Keep an eye on the bottom of the tape guide where the wrinkle is. At some point, the video tape will move up the tape guide and smooth out.

If you turn the tilt adjustment too fast, the video tape could pop up across the tape guide and position itself too high. If you overadjust the tilt, the video tape could start wrinkling at the top of the tape guide. If you have the TV monitor on at this time, watch the picture being produced from the video tape. You'll see a horizontal belt of lines across the screen. This belt of lines starts at the bottom of the screen and moves up towards the top of the screen and then starts over again. You can correct these two problems by turning the tilt adjustment screw clockwise slowly until the bottom edge of the video tape is just touching the bottom protruding edge on the tape guide, as shown in Fig. 10-4. Now proceed to the section in chapter 22, on "Tilt adjustment."

Do not touch any other adjustments on the A/C head. If you do, you can lose the audio and video syncrinization. If you accidently turn the wrong screw and one of these problems exists, proceed to chapter 22, "Audio head alignment."

If this doesn't correct the problem, there's another cause. Usually the problem results from the take-up spindle pulling too hard on the video tape. The problem is a bad take-up clutch assembly. Refer to chapter 28, "Take-up spindle problems," for the method of replacement.

10-2 Second location at which the video tape can be eaten.

Reasons why a VCR will eat a video tape

There are several reasons why your VCR will suddenly decide to eat one of your video tapes. These reasons range from residue buildup to a bad pinch roller. This section will show you how to solve these problems quickly and easily.

First reason

I've taken a capstan shaft assembly out of a unit to show you what to look for. Look at the upper and lower portions of the capstan shaft, as shown in Fig. 10-5. You're looking for residue buildup. Residue buildup keeps the pinch roller from setting flush against the capstan shaft. The pressure on the pinch roller is incorrect and causes the video tape to be pulled up or down the shaft and to be eaten.

Look at the video tape as it passes between the capstan shaft and pinch roller while the unit is playing. The tape is being crinkled up and eaten (Fig. 10-6).

To avoid this problem clean off the residue buildup on the capstan shaft. Immerse a chamois stick or glass brush into the cleaning solution. Once it's fully saturated, clean the capstan shaft, as shown in Fig. 10-7. Apply pressure with the cleaning utensil; the residue buildup can be quite hard. To clean the pinch roller, refer to the section in chapter 4, on "Pinch roller" under the subhead "Cleaning the various components."

10-3 Locating the tilt adjustment on the A/C head.

Second reason

The bracket that holds the pinch roller also can cause the video tape to be eaten. This bracket can be bent or twisted, causing uneven pressure between the pinch roller and the capstan shaft. To check for this condition, get a sheet of plain white paper and place the paper behind the pinch roller and capstan shaft, as shown in Fig. 10-8. Move the pinch roller closer to the capstan shaft. In most models, you can move the pinch roller over by pushing on the opposite end of the bracket the pinch roller is attached to (Fig. 10-9). In other models, push on the bracket right under the pinch roller to move it. Push the bracket over, leaving a thin gap between the two pieces. You should be able to see a thin white line between the two parts, as shown in Fig. 10-10. Check to see if the white line is the same width at the bottom as at the top. If it isn't, the bracket holding the pinch roller is bent or twisted.

In some models, the white line method isn't possible. In these models, the pinch roller bracket is in a locked position, enabling you to move it by hand. In other models, the cassette holder blocks the view of the line. In either case, there is another way to check for a bent or twisted bracket.

Check for residue buildup on the capstan shaft or pinch roller. If it is clean, insert a video tape and push Play. Observe the video tape as it goes between the capstan shaft and the pinch roller. See if the video tape is being pulled down and is crinkling at the bottom edge of the tape. If so, take a small screwdriver and place it at the top edge of the pinch roller cap, as shown in Fig. 10-11. Apply a small amount of pressure, pushing away from the capstan

10-4   Aligning the video tape to the take-up tape guide.

shaft. Do not push too hard or the pinch roller will stop turning. This operation should cause the video tape to straighten out between the pinch roller and the capstan shaft, but only when the pressure is being applied. On the other hand, if the video tape is being pulled up and is crinkling at the top of the video tape, put pressure towards the capstan shaft (Fig. 10-12).

Some models don't have a pinch roller cap. To apply the pressure, take a Phillips head screwdriver and place it in to the grooves of the screw on top the pinch roller. Then you're able to push the pinch roller with the shaft of the screwdriver to check the crinkle in the tape. This procedure should straighten out the video tape. In either case, if the video tape straightens out while pressure is being applied, then the pinch roller bracket has been bent or twisted. The take-up spindle also can cause this problem. Refer back to the last two paragraph in the section in this chapter entitled "Second location."

To remove the pinch roller bracket, remove the C- or E-ring, and pull the bracket straight up, and off, as shown in Fig. 10-13. For removal of C- or E-rings, refer back to the section in chapter 7 on "Removing C- or E-rings" under the subhead "Finding the lubricating points." In some models, the bracket is made of a thin metal and bends easily. Although the bracket varies in size and shape from model to model, you remove them all the same. The point at which it might get bent or twisted is shown in Fig. 10-14. You have an option. You can go to a local electronics supply center and purchase a new one, or you can straighten out the one you already have. Take the new or straightened bracket and place it in its original position. If you straighten out the bracket, recheck the white line, making sure the pinch roller is straight up and down before inserting a video cassette.

In some models, there might be additional parts to be removed, such as a mounting bracket or a small circuit board. These parts are held on with a screw. Simply take out the

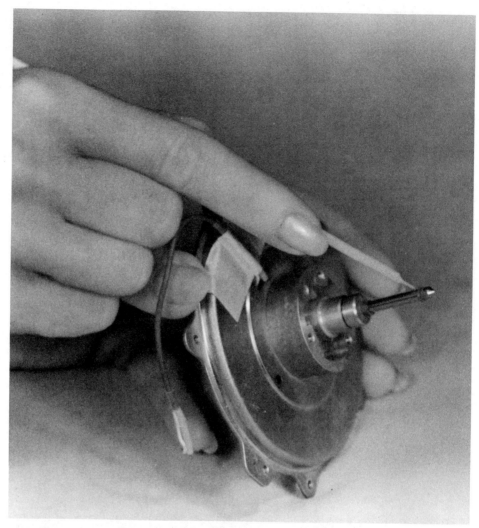

10-5 Locating residue buildup on the capstan shaft.

screw (Fig. 10-15) and remove the mounting bracket or the small circuit board, as shown in Fig. 10-16.

Another part you might find in the way is a spring. If so, take a stiff piece of wire or a paper clip and make a hook on the end of it. Place the hook through the hook on the end of the spring and pull the spring off the mount (Fig. 10-17). Place the spring out of the way.

In some models, the bracket might have another arm attached to it. You'll need to detach one end of the arm to remove the pinch roller bracket. All attached arms are held on by a C-or E-ring. Remove the ring, and the attached arm will come up and off, as shown in Fig. 10-18.

After removing all the parts that were in the way, remove the C-or E-ring that is holding down the pinch roller bracket. Take hold of the pinch roller and pull straight up and off, as shown in Fig. 10-19.

10-6   Video tape being eaten between the pinch roller and capstan shaft.

10-7   Cleaning the capstan shaft.

10-8 Placing a sheet of paper behind the pinch roller and capstan shaft.

10-9 Moving the pinch roller over to the capstan shaft.

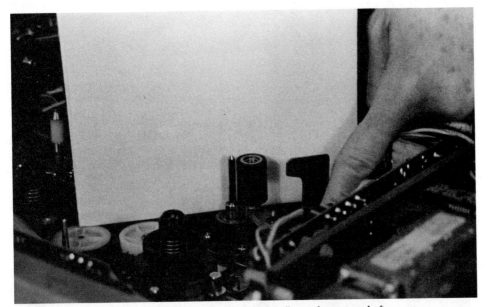

10-10   The gap between the pinch roller and capstan shaft.

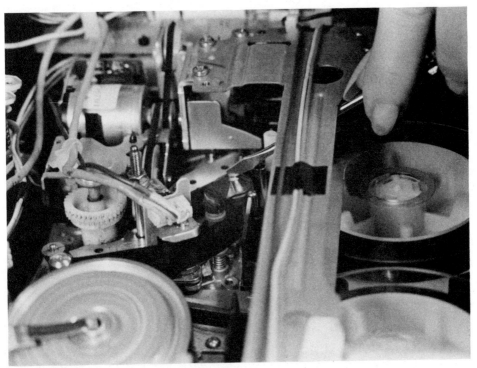

10-11   Pushing the top center of the pinch roller toward the back of the VCR.

10-12   Pushing the top center of the pinch roller toward the front of the VCR.

10-13   Removing the pinch roller bracket.

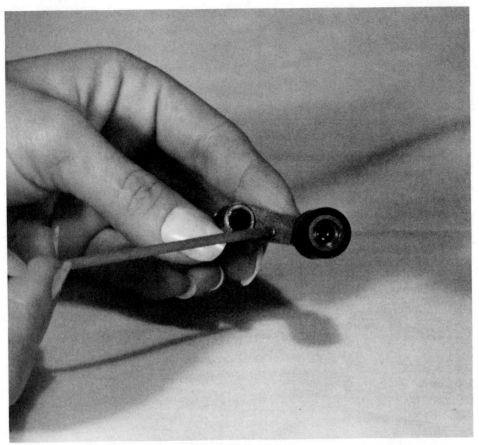

10-14   The point the pinch roller might get bent or twisted.

Third reason

There are two ways a pinch roller can go bad. The pinch roller might become hard and shiny like a piece of hard plastic, instead of hard rubber. The video tape will become unstable, move up and down on the capstan shaft, and be eaten. In this case, replace the pinch roller. Refer to the section in chapter 7 on "Removing a pinch roller" under the subhead "Finding the lubricating points." The pinch roller also can become warped, causing uneven pressure on the video tape. The fluctuating pressure causes the tape to rise or fall on the capstan shaft and to be eaten. To check for this type of problem, use the white line method, as previously described, and check for a bow in the line, as shown in Fig. 10-20.

You also can check the pinch roller by removing it from its bracket. Place the roller up against the shaft of a screwdriver and see if the side of the pinch roller has a bow in it. If so, take it to a local electronics supply center and give them the make and model number of the VCR for replacement.

Fourth reason

The pinch roller or capstan shaft can eat a tape if you insert a video tape with a sticky substance on it. This sticky substance is usually caused by a spilled beverage or some other

10-15 A circuit board obstructing the pinch roller mount.

10-16 Removing the circuit board or bracket.

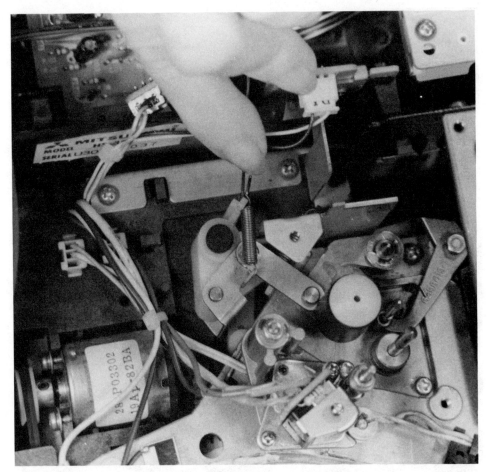

10-17   Pulling the spring off its mount.

substance contacting the video tape. Due to the pressure between the pinch roller and the capstan shaft, the video tape will ball up around the pinch roller or the capstan shaft. (Refer to "Seventh reason" under the heading "Reasons why the cassette holder jams" in chapter 9.) If you find a ball of video tape around the capstan shaft, remove it. Thoroughly clean both parts as previously explained.

Fifth reason

There's a video cassette in the unit, and you push Eject. When the cassette comes out, it leaves about a foot or so of video tape stuck inside the unit. The tape might come out with the cassette, but it's crinkled. The problem is that the supply spindle didn't pull the video tape back into the cassette, when the VCR left the play or record mode. To correct this problem, refer to chapter 29, "Fast forward and rewind problems." The faulty part is probably the idler wheel tire. This part drives the supply spindle.

10-18 Removing the attached arm from its bracket.

Review

☐ Examine the following when looking for problems with the pinch roller and capstan shaft.
  • Residue buildup on capstan shaft.
  • A bent or twisted pinch roller bracket.
  • A bow in the pinch roller.
  • The rubber on the pinch roller.
  • A ball of video tape around the capstan shaft.

☐ If you find a ball of video tape on the capstan shaft, it's not the fault of the VCR, but a sticky substance that was left on the video tape.

☐ When the tape is being wrinkled at the bottom of the take-up tape guide, do not touch the adjustment nut on top of the tape guide. Replace the take-up clutch assembly or turn the A/C head tilt adjustment.

☐ When adjusting the A/C head, make sure the video tape is flat against the tape guide and the tilt adjustment is turned as far clockwise as possible without wrinkling the video tape.

10-19   Removing the pinch roller and bracket.

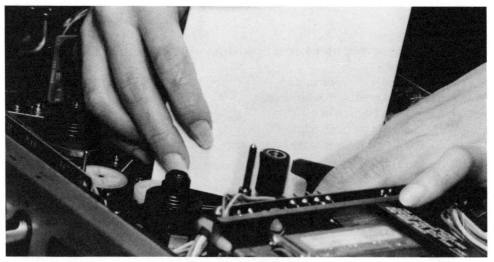

10-20   A bowed pinch roller.

11

Repairing a
video tape

ACCORDING TO MURPHY'S LAW, IT NEVER FAILS. SOMEHOW THE SPECIAL VIDEO tape of your child's birthday party or Uncle George's 75th anniversary has been eaten by the VCR.

What usually happens to a video tape
- The capstan shaft crinkles up the video tape.
- The video tape sticks to the capstan shaft or pinch roller, wading it up into a ball.
- You break the video tape, trying to remove it from the VCR.
- One of the broken ends gets rewound back into the video cassette where you can't reach it.

To remedy all of these dilemmas carefully follow the following instructions. It's important to remember that grease, dirt, or any oil on your hands can rub off onto the video tape. Wash and dry your hands thoroughly before touching the video tape.

Opening a video cassette

Start by facing the front of the video cassette towards you. You'll find a small button on the left side, just behind the door. Push this button to release the door on the front of the cassette. Pull the door open all the way. Take a piece of scotch tape and tape the door back, as shown in Fig. 11–1. Flip the cassette over and remove the five screws on the bottom of the video cassette (Fig. 11-2). Check the sides and back of the video cassette for any labels. If the labels have been on the cassette for a long period of time, they might be difficult to get off. Take a single-edge razor blade and slice each label down the middle, following the crack along the sides and back of the video cassette, as shown in Fig. 11-3. Hold both the top and bottom covers at the same time. Turn the cassette right-side up and lift off the top cover (Fig. 11-4).

11-1 Taping back the door of the video cassette.

11-2 The five screws on the bottom of the video cassette.

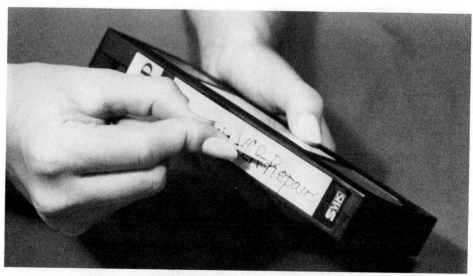

11-3   Slicing the label with a razor blade.

11-4   Lifting off the top cover.

Unlocking the reels

After removing the top cover, you can see two exposed reels inside the cassette, as shown in Fig. 11-5. When looking at each reel, notice the teeth on the outside of each reel. These teeth are locked down by lock arms to keep the reels from turning when the cassette is inactive. To release the reels, pull the lock arms away from the teeth, allowing the reels to turn freely.

11-5   The reels and locking arms inside the cassette.

Threading the video tape

Face the video cassette towards you. This side the video tape broke off, and pulled up inside the cassette (Fig. 11-5). Take hold of the take-up reel, which is the reel on the left, and pull it straight up and out. Now, pull out a foot or so of the broken video tape, as shown in Fig. 11-6. Lower the reel close to the cassette and place the pulled out video tape between the white plastic roller guide and the metal guide (Fig. 11-7). Now lower the reel back down into its original position.

Now, look at the supply reel, the reel on the right side of the cassette. On this side, the video tape has been eaten and crinkled. Release the lock arm and pull out some good video tape. Take a look at the roller guides on this side. The video tape goes between a metal pin and a fiber spring, then around a metal guide. In some models, the metal guides have a white plastic roller around them. In Fig. 11-8, the fiber spring is being pulled away to give you a better view.

Splicing a video tape

Cut off all the crinkled up video tape, as shown in Fig. 11-9, leaving only smooth video tape coming out from each reel. Place the two smooth ends of video tape, overlapping them by about three inches. Make sure each end of the video tape is perfectly aligned. In other words, both pieces should be laying on top of each other, appearing as one piece. Make sure the video tape coming from each reel isn't twisted. Take a pair of sharp scissors and cut both pieces of video tape at a slight angle (Fig. 11-10). By cutting the video tape in this manner, the ends will match up perfectly.

Get a roll of ¾-inch-wide scotch tape and tear off a piece approximately two inches in

11-6   Pulling the video tape out of the reel.

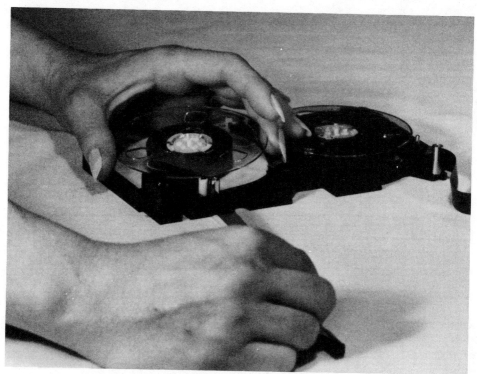

11-7   Placing the video tape between the take-up guides.

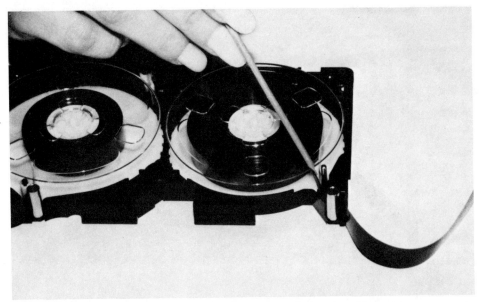

11-8   Placing the video tape between the supply guides.

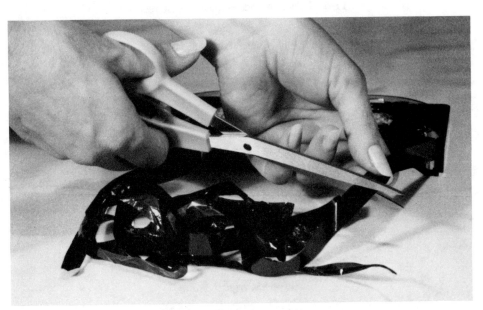

11-9   Cutting off the crinkled video tape.

length. Hold the piece of scotch tape in one hand with the sticky side up. Take one of the ends of the video tape and place the side that will not come in contact with the video heads approximately two-thirds of the way across the scotch tape. Attach the video tape to the scotch tape at this point, as shown in Fig. 11-11. Now lay down the two attached pieces of

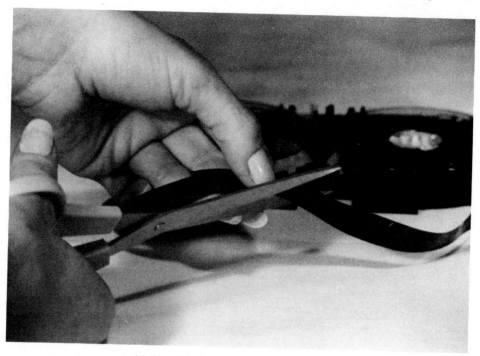

11-10 Aligning and cutting the video tape.

tape with the sticky side up. Take the other end of video tape coming out of the video cassette and place the backside of it on top of the taped piece laying on the table. Be careful. Do not let the top piece of video tape touch the scotch tape yet. Reposition the video tape so both sides are perfectly aligned, as shown in Fig. 11-12. Next, gently slide the top piece of video tape toward the cut end of the bottom piece until the top piece drops off the end.

If you did everything right, both ends should be perfectly aligned and appear as a single unbroken piece of video tape (Fig. 11-13). Take your finger and press down on the video tape to secure it to the scotch tape. If the second piece of video tape should stick to the scotch tape in the wrong place, don't try to remove it. Cut the taped section out and start over.

Trim off the excess scotch tape. Use a sharp pair of scissors to cut a hairline off the edge of the video tape, while cutting off the scotch tape. Notice the fine line of video tape stuck to the scotch tape in Fig. 11-14. Cut the other side the same way. By cutting the scotch tape in this manner, you won't leave any exposed to get caught in the video heads.

Testing the splice

Take a clean finger and press down on the front side of the video tape over the splice, as shown in Fig. 11-15. If the video tape sticks to your finger, the sticky part of the scotch tape is exposed and will clog the video heads. Cut this section out and start over. You have to be sure there's no exposed scotch tape. It might take you two or three times before you get the splice right.

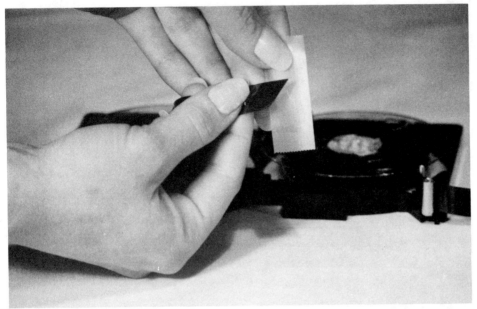

11-11 Attaching the video tape to the scotch tape.

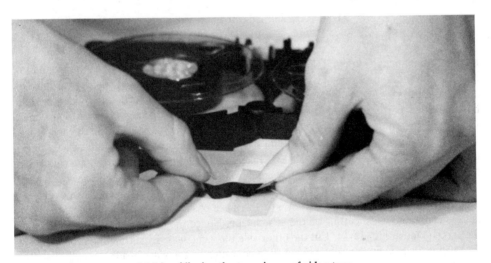

11-12 Aligning the two pieces of video tape.

Reassembling the video cassette

Turn either reel to pull the excess video tape back into the cassette, as shown in Fig. 11-16. Now, you are ready to put the video cassette back together. Align both tape guide pins on the bottom cover to the matching holes on the top cover. (See Fig. 11-17) Push the two covers together. Flip the cassette over, put the screws back in their proper holes, and tighten

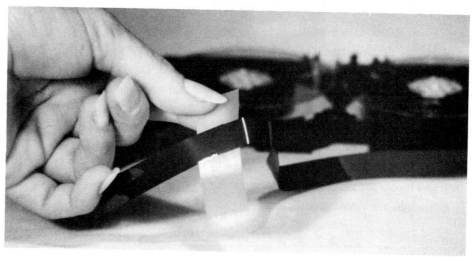

11-13   A perfectly aligned splice.

11-14   Leaving a fine line of video tape on the scotch tape.

down each screw. Remove the scotch tape to release the cassette door, and the video cassette is ready to go.

Review

☐ Open the cassette door and tape it back.

☐ Take out the five screws located on the back of the cassette.

☐ Flip the cassette over and remove the top cover.

☐ Be sure to wash hands before touching the video tape.

☐ Release the lock arms to pull out the video tape.

☐ Make sure the video tape is properly threaded through the roller guides.

☐ Cut the wrinkled tape off.

☐ Cut both ends of the video tape at a slight angle and at the same time, for a perfect match.

☐ Using scotch tape, carefully tape both ends of the video tape together.

☐ It's important to place the scotch tape only on the backside of the video tape.

☐ Check the splice for any exposed scotch tape by using a clean finger.

☐ Turn a reel to pull the excess tape back into the cassette.

☐ Reassemble the video cassette.

11-15   Testing the splice for exposed scotch tape.

11-16 Pulling the excess tape back into the cassette.

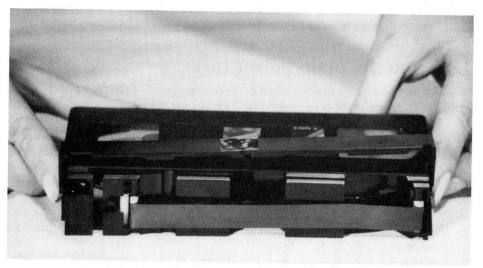

11-17 Aligning the video cassette pins.

12
Replacing fuses

THERE ARE FROM ONE TO SIX FUSES INSIDE A VCR. MOST FUSES SNAP INTO A holder and are easy to replace. A few older models have fuse holders with twist off tops. These fuses also are easy to replace. Before proceeding with this lesson, unplug the unit. Be sure to read the section in chapter 2 on "Electrical shock."

Locating the main power fuse

If the VCR goes completely dead, nothing will light up, including the digital clock. If this situation occurs, check the main power fuse. First remove the top main cover. Locate the main fuse by following the ac cord to where it enters into the back of the unit. From there, the ac cord goes either into a metal plastic covered box or onto a circuit board.

Older models use a metal box. Usually a twist top fuse mount is on top of the metal box, as shown in Fig. 12-1. To remove the fuse, place a flathead screwdriver into the slot on top of the fuse mount. Turn the slot counterclockwise until the top of the fuse holder comes off. Pull the top of the fuse holder straight up, and the fuse will come out with it. (see Fig. 12-2)

Some VCRs use a clear plastic covered box to hold the fuses. You can see the fuse inside the box, as shown in Fig. 12-3. Simply unsnap the plastic cover and pull it to the side to remove the fuse.

There also are opaque plastic boxes. In this case, follow the ac cord into the unit to where it enters the plastic box (Figure 12-4).

There might be mounting screws on each side at the base of the plastic box. Pull the plastic cover straight up and off, as shown in Fig. 12-5. You'll find a small circuit board under the plastic and with the main fuse mounted to it (Figure 12-6).

For models that use a circuit board to hold the fuses, follow the ac cord to where it connects to the circuit board. Right beside the connection is a snap-in fuse. This fuse is the main power fuse. You might find only one fuse mounted to the circuit board, but there can be

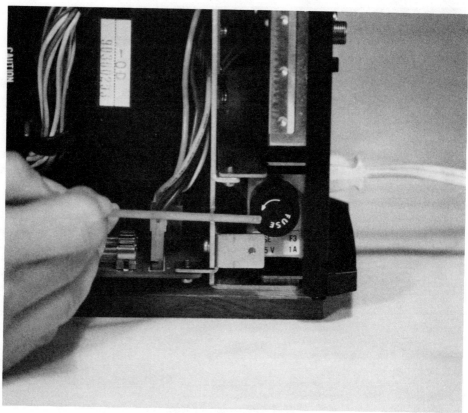

12-1 A twist-type fuse holder.

up to five fuses in a row on the same circuit board. The main power fuse is closest to the ac cord connection, as shown in Fig. 12-7.

To remove a snap-in fuse, place a small flathead screwdriver under one end of the fuse, and pry the fuse up, using the circuit board as a lever, as shown in Fig. 12-8.

Checking a fuse

There are two ways that a fuse can go bad. If a fuse blows out, it turns black inside the glass tube, as shown in Fig. 12-9. This condition usually means the VCR has a dead short. The second way is that the fuse opens up, that the thin wire inside the glass tube has broken. An open fuse is usually caused by a powder surge in the unit. Sometimes it's hard to tell if the wire is broken. To test for a broken wire, remove the fuse and tap the glass tube. If the wire inside starts vibrating, the fuse is bad. It's possible for the fuse to look good but to have a bad connection inside.

Some fuses have a white coating inside the glass. It's not possible to check inside this type of fuse to see if it's blown. Take the fuse into a local service center and have the fuse checked.

If the main power fuse has turned black inside, the power supply probably has shorted out. If you replace the fuse, it will just blow out again. In this case, you should take the unit

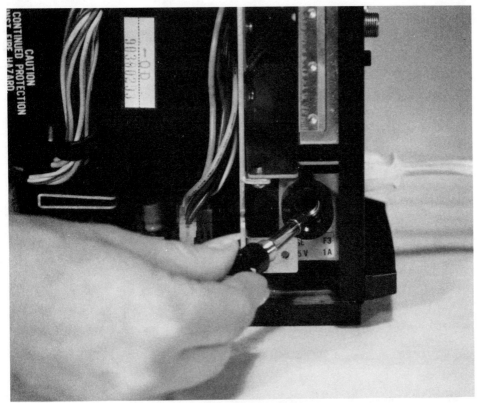

12-2   Removing the fuse.

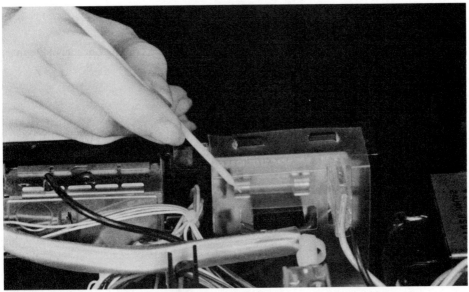

12-3   Locating the main power fuse in a transparent covered box.

12-4   Locating the main power fuse in an opaque covered box.

to a local service center for circuit repairs. If the fuse has opened up, you probably had a power surge, and all that's wrong is an open fuse. You can go ahead and replace the fuse.

Locating secondary fuses

Look at the rest of the fuses in the VCR. You might find one to five more fuses inside, in addition to the main power fuse. These fuses usually are near the power transformer, on the component side of a circuit board. These fuses run on a low voltage, and each fuse supplies a separate portion of the unit. If a certain function stops working but the rest of the unit is functioning properly, check the secondary fuses. If you find a blown fuse, you have a circuit problem. If the fuse has just opened up, replacing the fuse will probably fix the problem.

Reading the value of a fuse

To read the value of a fuse, look for printing on each metallic end of the fuse. For example, on one end of the fuse it reads 125V5A. This label means the fuse has a voltage of 125 volts and an amperage of 5 amps. Voltage is a measurement of electricity. Amperage is the strength of the electrical current. You read all fuses the same.

Review

☐ Always unplug the unit before replacing a fuse.

☐ There are always one to six fuses inside a VCR.

☐ To find the main power fuse, always follow the A/C cord to a covered box or to a circuit board.

☐ A blown fuse turns black, and an open fuse just breaks the wire inside the glass.

☐ All secondary fuses are located on a circuit board close to the power transformer.

12-5   Pulling off the plastic cover.

12-6 Locating the main power fuse on a circuit board.

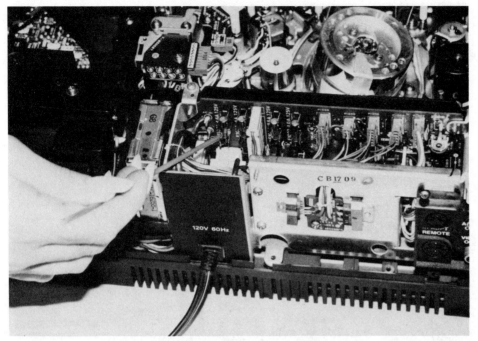

12-7   Locating the main power fuse.

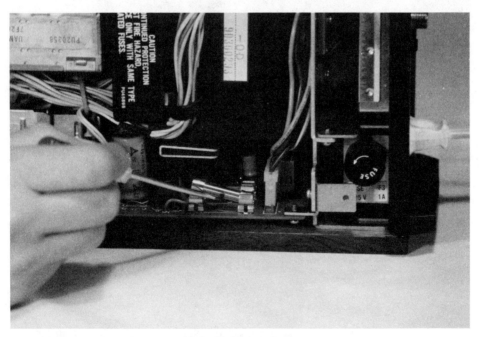

12-8   Prying out the fuse.

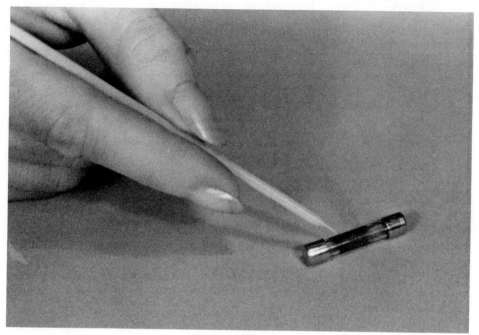

12-9 A bad fuse.

13

Repairing water
and beverage damage

LIQUID IS ACCIDENTALLY SPILLED ONTO THE VCR, AND SOME RUNS DOWN
inside the unit. If the liquid lands on any metal or plastic parts or on the cassette carriage
itself, the unit will keep on running like nothing happened. It still is important to immediately
unplug the VCR, remove the main top cover, and wipe up the liquid inside the unit. Cleaning
up the liquid will keep any metal from rusting.

Water damage

Water is a conductor of electricity; if water lands on a circuit board, it will short out the board.
The board will stop functioning, and any functions performed by the board will not work. If
spilled water causes a board to short out, immediately unplug the VCR, remove the main top
cover, and wipe up the water. You can use a blow dryer to completely dry out the unit and
circuit boards.

If you don't have a blow dryer, wait a day or two and then recheck the circuit board to
see if the water has completely evaporated. If the circuit board has a lot of dust or dirt on it,
the water will mix with the dirt, forming a mudlike substance. Because of this substance, it
can be hard to tell if the water has completely evaporated. If the circuit board is clean and the
water has completely evaporated, plug in the unit and turn it on. When water comes in contact with a circuit board, the water usually just grounds everything out. When the water
completely evaporates, the circuitry goes back to normal, and the unit starts working again.

If the unit doesn't start working, check all the fuses. Sometimes the water will land in
just the right spot and blow a fuse. Replace any blown fuses. After the mudlike substance
dries, it turns into a mud stain (Fig. 13-1). Clean the circuit board off before you plug in the
unit.

Place the unit on its side and thoroughly spray the mud stained area with degreaser. The
degreaser might not remove the entire stain. If the stain is not completely removed, spray

116

13-1 A mud stain on a circuit board.

an old toothbrush with degreaser. Brush the stain vigorously, as shown in Fig. 13-2. Follow the same procedure for any other stained areas. Respray the interior circuit board with degreaser to remove any extra substance left on the board.

In Fig. 13-3, the mud stain is on the circuit board, around and under all the components. Some of these components, such as the capacitors and transistors, can be moved out of the way. A capacitor is a round cylinder with two wires attached to the circuit board. A transistor is a small black rectangular box with three wires attached to the circuit board. Usually these two types of parts are mounted about ½ inch away from the circuit board on thin wires that bend easily. Gently move these components out of the way, as shown in Fig. 13-4.

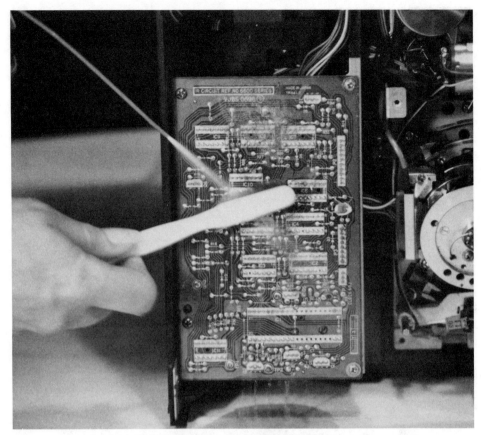

13-2   Scrubbing and spraying with degreaser.

There are other components, such as resistors, and ICs, that are tightly mounted to the circuit board and can't be moved. A resistor is a long cylinder mounted parallel to the circuit board, with a wire coming out of each end. An IC is a black rectangular box with many pins coming out of each side. It has the appearance of a centipede and is soldered directly to the circuit board. Spray the stained area with degreaser. Use a toothbrush to scrub the circuit board, including the top and around each part, until the board is clean. (See Fig. 13-5) Respray the entire circuit board to remove any substance left on it. Finally, move each part back into its original position, being very careful not to short any wires together.

Beverage damage

Let's discuss beverages like beer, and soda pop. If you spill a liquid such as beer or soda pop, into the VCR, the liquid will leave a sticky coating that will take months to dry. Unplug the unit and locate the all stained areas. (Be sure to check all the nooks and crannies around the affected areas.) Clean all the affected areas in the VCR as explained in the last section. Check for any blown fuses.

13-3 A stained circuit board on the component side.

Review

☐ If any fluids have spilled into the VCR, unplug the unit immediately.

☐ Let the unit sit for a day or two, allowing the liquid to completely evaporate.

☐ If the damage is caused by water:
  - Immediately remove any excess water.
  - Check to see if the circuit board is clean.
  - Wait until the water has evaporated.
  - Plug the unit in. It should work. Replace a fuse if necessary.

☐ If you have a mudlike substance on a circuit board, you'll need to clean it off before you can plug in the unit.

☐ If you spill a beverage in the VCR, you'll need to clean all affected areas.

☐ Check all moving parts that could have come in contact with the beverage and clean them.

13-4 Moving a capacitor out of the way.

13-5 Scrubbing and spraying the component side.

14
Diagnosing VCR problems

THIS CHAPTER COVERS THE PROCEDURES FOR DIAGNOSING PROBLEMS IN A VCR. The problems have been grouped into three major categories based upon the types of symptoms they exhibit: video, audio, or mechanical. You'll be referred to the proper chapter for instructions on how to correct each particular problem.

Video

Many of the problems that your VCR will develop will be noticed because of the effects they will have on the picture that you see on your television. Choose the section that describes the problem you are experiencing and follow the instructions describing how to repair the problem.

Distorted picture

While you're viewing a video tape, the picture becomes partially or completely distorted or snowy. The TV screen might even go completely black or blue; however, the audio portion is okay. If you have a stereo hi-fi unit, place the stereo switch to the normal position. Listen to hear if the audio clears up. Stop the VCR and tune it to a broadcast TV station. If the program comes in clear on the TV monitor, you could have a dirty video head. Refer back to chapters 5 and 6, "Cleaning the video heads" and "Pinpointing and correcting video head problems." If ½ to ⅔ of the screen is distorted or snowy, but the remainder of the screen is clear, proceed to chapter 20, "Video tape path alignment."

Lines following an object

You see black or white trailing lines shooting off the right side of an object in the picture. They are on the entire screen. The more pronounced the object, the more visible the lines. The cause of this problem is a bad video head. If trailing lines still appear after cleaning and saving the video head, then the video head is bad and has to be replaced.

If the lines can be adjusted out of the picture with the tracking control, the video head is on the borderline of going bad. Refer to the sections in chapter 19 on "Adjusting the back tension guide" and "The brake shoe band adjustment." Then refer to the section in chapter 23 on "Centering the tracking control."

Stationary lines on the top or bottom of the picture

When you play back any video tape, you find a group of horizontal lines across the top or bottom of the screen. These groups of lines are stationary. (They do not float through the picture.) You also will find the tracking adjustment moves these lines up or down the screen, but they still remain on the screen. You can have many groups of horizontal lines across the screen, covering nearly the entire picture. If either of these conditions exist, the video tape path is out of alignment. Proceed to Chapter 20, "Video tape path alignment."

Lines appearing in the picture on some tapes

If you have to adjust the tracking control everytime you play a video tape to remove the lines from the picture, then you need to recenter the tracking control. Proceed to Chapter 23, "Adjusting the tracking control." If lines resembling those that appear in fast-forward search are running through the picture when you play a six-hour tape, proceed to chapter 22, "Audio head alignment."

Jitter in the picture

You're viewing a video tape and there's a jitter in the picture, which means the picture jumps up, and down. You adjust the tracking control on the VCR and check the vertical hold control on the TV monitor, but these adjustments have little or no effect. Proceed to the sections in chapter 20 on "Checking the adjustment" and "Fine-tuning the roller guides."

You also should check the torque on the video tape. If the picture is okay in the two-hour mode but has a jitter in the six-hour mode, proceed to the section in chapter 19 on "Adjusting the back tension guide." Adjust the back tension guide for a six-hour video tape.

Lines floating through the picture

You're viewing a video tape, and you see a belt of lines moving continuously from the bottom to the top of the screen. The lines also can be floating from top, and bottom and be meeting in the middle. This condition usually is caused by the FM portion of the A/C head being out of alignment. To check for this problem, use a fiberglass brush to push the bottom portion of video tape up against the A/C head while the unit is playing, as shown in Fig. 14-1. Using the brush, move the video tape up slightly. If moving the tape up doesn't clear the picture, then push the video tape down slightly. If either position clears the picture, then do an audio head alignment. This adjustment automatically aligns the FM portion of the A/C head, which removes all the lines. Proceed to chapter 22, "Audio head alignment," and read "A tip on adjusting the FM portion."

Another way to diagnose this problem is to place your finger on top of the back tension guide pole, then push the guide pole to the left while the unit is playing. If this adjustment clears the picture, proceed to chapter 19, "Setting the torque on a video tape." Either the back tension isn't tight enough or the audio head is out of alignment. If neither of these adjustments clears the picture, you have a bad A/C head or a circuit problem. You'll have to take your VCR in for service.

14-1 Pushing the bottom portion of the video tape against the A/C head with a glass brush.

Picture jumping up and down

There is a small group of lines across the bottom portion of the screen, or the whole picture jumps up and down. The vertical hold control on the TV monitor and the tracking control on the VCR won't remedy either condition. Stop the VCR and eject the video cassette. Locate the button on the front left side of the video cassette. Push this button to open the cassette door, exposing the video tape. At the bottom edge, you'll see a crinkle (Fig. 14-2), which was caused from the tape previously being eaten. If this problem exists, refer to the section in chapter 10 on "Second Location" under the subhead "Two locations at which a video tape can be eaten."

Lines and dashes across the screen

You see many stationary lines and dashes shooting across the screen. The audio portion is distorted or has a buzz in it. This problem is caused by the back tension guide pole being out of adjustment. Proceed to chapter 19, "Setting the torque on a video tape."

You see a single long white or black line shooting across the screen like a star. There also might be one line on the top, middle, or bottom of the screen at the same time. The lines flash sporatically anywhere on the screen, then disappear. This condition, called video drop

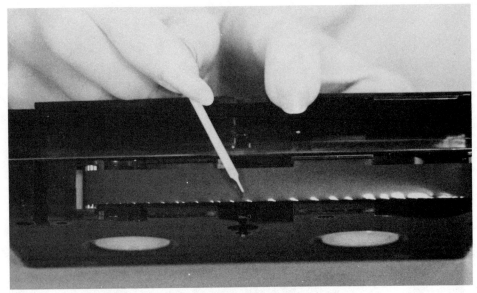

14-2 A crinkle on the bottom portion of the video tape.

out, is usually caused by a video head going bad. Refer to chapter 5, "Cleaning the video head," and chapter 6, "Pinpointing and correcting video head problems."

If you can adjust the lines out of the picture at either end of the tracking control, the FM adjustment is out of alignment. Refer to the sections in chapter 23 on "Locating the FM adjustment" and "Centering the tracking control."

Fluctuating distortion in the picture

The picture fluctuates from being stable and clear to being unstable, with a distorted picture. Each cycle takes 5 to 15 seconds. The audio portion sounds like it's speeding up and then slowing down. The audio portion also might be distorted and then clear up simultaneously with the picture. Proceed to chapter 19, "Setting the torque on a video tape," or to the section in chapter 22 on "A tip on adjusting the FM portion." The FM synchronization on the video tape isn't aligned with the A/C head properly, causing this problem.

Multiple floating lines

Periodically, you might see various size of lines floating through the picture. Along with this symptom, you'll hear a buzz in the audio. If this condition occurs while viewing a particular tape, stop the VCR immediately. Remove the video cassette, open the cassette door, and look at the tape. In this particular case, you'll find the video tape has been eaten across its whole width, as shown in Fig. 14-3. Refer to chapter 10, "Video-tape-eating VCRs."

A bar of white rectangular and square boxes

You see a configuration of white rectangular and square boxes that resembles a large city. The configuration is a long, narrow bar across the screen. It repeatedly starts at the top of the screen, works its way down, and disappears. It also might remain stationary at the top of

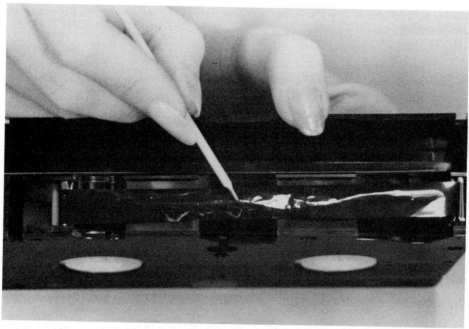

14-3 A video tape eaten completely across.

the screen, appearing and disappearing. Refer to the section in chapter 19 on "Detecting a back tension guide problem," or chapter 5, "cleaning the video heads."

A few floating lines and then the vcr stops

The VCR is playing, a few lines suddenly appear on the screen, then the unit automatically stops. It repeats the process if you push Play again. If the problem persists, push Fast Forward for a second or two, push Stop, and again push Play. The VCR stops at the same point every time because of a small pinhole in the video tape. When the pinhole passes the tape sensors, the sensors signal the unit to stop. In this case, the problem isn't the VCR, but in the video tape.

Speed fluctuation in playback causing lines

You have recorded a program from a broadcasted TV station. You believe it recorded just fine, but upon playback of the recording, you see and hear the picture and audio fluctuating. This problem isn't caused by the VCR, rather by the antenna system interfering with the TV signal. A distorted cable system also can cause this interference. The interference blocks the internal speed governor signal, which determines the proper tape speed. The interference causes the unit to continuously search for the right speed during playback, causing the unit to speed up and slow down.

I've found that many people use improper antenna systems. The worst antenna system is rabbit ears. This type of antenna picks up interference, even from someone walking too close to it, causing the TV signal to be weak or ghosts to appear in the picture. A weak signal means that only a weak sync pulse reaches the unit; consequently, the unit can't determine at what speed it should run. The following outside antenna systems cause the same interference as rabbit ears:

☐ An antenna with broken or bent arms.
☐ An antenna that is not attached properly and flops around.
☐ An antenna aimed in the wrong direction.
☐ Using bad lead-in wire.
☐ Using a plain piece of wire for an antenna.
☐ A weak cable system.

Audio

If the audio portion of your video tapes begins to sound strange, the problem can be traced to one of only a couple of possible causes. Read the sections below and follow the instructions to repair the problem that you are having.

Distorted audio

Upon playback, the VCR is producing a good picture, but the audio is distorted or is coming in very weak. Here are five explanations why this situation can happen. A buildup of residue on the face of the audio head from dirty video tapes can cause audio distortion. This buildup prevents the video tape from making a flush contact with the audio head, giving a weak or distorted audio signal. Refer to the section in chapter 4 on "A/C head" under the subheading "Cleaning the various components." Improper alignment is another cause of audio distortion. Proceed to chapter 22, "Audio head alignment." The third cause is a bad audio head or a circuit problem. Proceed to the section in chapter 24 on "Checking for a bad audio head." The last two causes occur only in stereo hi-fi models. If the unit is in the stereo hi-fi mode and the audio is distorted or makes a motorboat sound, this distortion might be caused by the tracking control being set improperly. To remedy this complication, readjust the tracking control until the audio clears up. Another reason stereo hi-fi units might have distortion, is the video tape path being out of alignment. Proceed to chapter 20, "Video tape path alignment." The audio in these types of models is produced by the video heads. If the tape path is out of alignment, the audio is distorted or has a buzzing sound in the background. If the audio is clear on the stereo hi-fi mode but is distorted in the normal position, check for a bad or dirty audio head.

Wavering in the audio

You hear a wavering in the audio when playing a video tape. It's more noticeable in the six-hour mode than the two-hour. You also hear this wavering in all rented video tapes. You'll notice that, if you play the recording on another VCR, the wavering sound is still there. The problem is caused by a bad clutch assembly. Proceed to the section in chapter 28 on "Removing the take-up clutch assembly" and the section in chapter 29 on "Removing idler arms and clutch assemblies." The type of clutch that usually causes this problem resembles two gears sandwiched together and is located in the undercarriage. Purchase a new clutch assembly at a local electronics supply center. Take in the make and model number of the VCR.

Mechanical

The final category of VCR repairs is mechanical problems. These problems don't display symptoms on the television, visually or audibly. Look through the following sections, find the heading that describes your problem, and follow the instructions in that section.

VCR shutting down

The unit completely shuts down with a video cassette inside. You've tried pushing Power, but there's no effect. Try unplugging the unit and plugging it back in. The microcomputer will reset. When you plug the unit in, it'll fire up right away or you'll have to push power. You'll hear a motor running or a loud squeal coming from inside the unit. This sound continues for about 60 seconds, then the unit shuts down. The cause of this problem is a bad loading belt. Refer to chapter 25, "Getting into the undercarriage," and to the first four sections of chapter 26. Plug the unit back in while it is on its side and the undercarriage is open. Check which motor is running. The belt attached to this motor has gone bad.

If the unit loads without a video cassette inserted then shuts down, unplug the unit and plug it back in. The cassette ejects after about 10 seconds. If the unit loads but shuts down again, look for a broken gear on the carriage. Refer to the section in chapter 9 called "Sixth Reason" under the subheading "Reasons why the cassette holder jams."

VCR stopping

The unit is in play, fast forward, or rewind. All three functions are performing properly. After 60 seconds or less the unit stops. The cause of this problem is that the counter belt is slipping or has come off the groove on the take-up spindle. Proceed to chapter 30, "Tape counter problems."

You've inserted a video tape, and pushed play. The unit runs for more than 60 seconds with a good picture, then stops. Each time you push Play, you receive the same results. The cause of this problem is that the take-up spindle isn't turning. Proceed to chapter 28, "Take-up Spindle Problems."

Clicking or grinding noise

A clicking sound coming from the unit itself is caused by the tape counter or by the small gear attached to the take-up spindle. Proceed to the section in chapter 29 on "VCRs using a direct drive system" or to chapter 30, "Tape counter problems."

In fast forward, you hear a grinding noise. The reels inside the video cassette aren't moving. The unit stays in this mode for about 60 seconds, then the unit stops. Proceed to the section in chapter 29 on "VCRs using a direct drive system."

Loud squeal

You've inserted a video tape and pushed Play. You hear the unit load up. Just before the picture should appear, you hear a loud squeal. The unit either stops or shuts down. Remove the main top cover, plug the unit in, and push Play. Watch the roller guides load the video tape to the video tape path. Notice that the roller guides don't quite reach the V-mounts when you hear this loud squeal. The video tape isn't being pulled into the tape path, then the unit stops or shuts down. The cause of this problem is a bad loading belt in the undercarriage. Refer to chapter 25, "Getting into the undercarriage," and the first four sections of chapter

26. After getting into the undercarriage, turn the unit on its side, plug the unit in, and push Play. Look for the motor that's running and the belt that's squealing. This belt is the one you need to replace. Refer to the sections in chapter 27 on "Loading motors" under the subhead "Identifying dc motors" and "Removing loading motors."

While playing, the unit has a good picture and sound, but you hear a squeal or screeching in the background coming from the unit. Sometimes the sound is present immediately after the unit starts to play or record, or it might appear even 20 minutes later. Proceed to chapter 34, "High-pitched sound coming from a VCR" or to chapter 31, "Capstan shaft problems."

VCR appearing to go in play

When you push Play, the unit starts up, but there is no picture or sound. After a moment the unit stops. Each time you push Play, the same thing occurs. All the other functions are working, including fast forward and rewind. The cause of this problem is a bad loading belt. Proceed to chapter 25, "Getting into the undercarriage" and to the first four sections in chapter 26. Then proceed to the sections in chapter 27 on "Loading motors" under the subhead "Identifying motors" and "Removing loading motors." Another cause could be the capstan shaft. Proceed to chapter 31, "Capstan shaft problems."

After inserting a video cassette and pushing play, the indicator light comes on, but the unit doesn't function in any mode. The indicator lights work as if the unit were functioning, but there's no effect. The cause of this problem is a bad loading motor. Refer to the sections in chapter 27 on "Loading motors" under the subhead "Identifying motors" and "Removing loading motors."

All mode buttons are nonfunctional

You've inserted a video tape, but pushing Play has no effect. You notice the Rewind, Fast Forward, and Record buttons also are nonfunctional. For the method for repairing this problem, proceed to chapter 32, "Cassette lamp problems."

Indicator lights flash off and on

There's a tape inside the unit, and the indicator lights are flashing off and on. All function buttons are nonfunctional. Try unplugging the unit and plugging it back in. The function lights should stay on for about one minute, then they'll start flashing again. The problem is a bad loading belt. Proceed to chapter 25, "Getting into the undercarriage" and to the first four sections of chapter 26. Then proceed to the sections in chapter 27 on "Loading motors" under the subhead "Identifying motors" and "Removing loading motors."

Stuck in fast forward

You turn the power on, and the unit goes directly into fast forward. The tape won't eject. Pushing all the mode buttons has no effect. Remove the main top cover, turn the power on, and view the take-up reel. If the take-up reel is spinning but the supply reel isn't, the video tape broke away from the supply spindle when rewinding. The tape wasn't properly secured to the reel. The sensors inside the unit signaled the unit to fast forward. To correct this problem, refer to the section in chapter 9 on "Manually removing video cassettes" under the subhead "A jammed video cassette in a topload VCR." After removing the video tape, the unit will start functioning properly. To repair the video tape, refer to chapter 11, "Repairing a video tape."

Stuck in pause

You've inserted a video tape and pushed Play. When the picture appears on the screen, it seems to be in pause. It'll stay in pause for 60 seconds and then stop. The cause of this problem is a bad tape drawer pin. A tape drawer pin holds the video tape away from the capstan shaft in fast forward or rewind. When in play, the drawer pin moves out of the way, allowing the pinch roller to meet the capstan shaft. At the opposite end of the bracket holding the tape drawer pin, there's another pin going into the undercarriage. The bracket is made of plastic. If it cracks, the pin going to the undercarriage can fall out. The bracket and the pin are shown in Fig. 14-4. If the pin is missing, the pinch roller can't go up against the capstan shaft; therefore, the capstan shaft can't pull the video tape through the video tape path. To correct this problem, locate the pin, which is located inside the undercarriage. Insert it back into the bracket and glue it into place.

Some models have no drawer pin. If you find the same symptoms but your unit has no drawer pin, the cause is a bad loading belt. Refer to the sections in chapter 27 on "Loading motors" under the subhead "Identifying motors" and "Removing loading motors." For instructions on cleaning the pulleys, refer to the first four sections of chapter 26.

14-4 The pin going into the undercarriage that can fall out.

A common recording problem

Countless numbers of times I've heard people say, "My VCR won't record anymore." If you have this problem, look on the back right corner of the video cassette for a missing tab (Fig. 14-5). This removable safety tab prevents the VCR from recording over a previously recorded video tape. If this tab is missing, cover the hole with a piece of electrical or scotch tape. You can now reinsert the video cassette and begin recording.

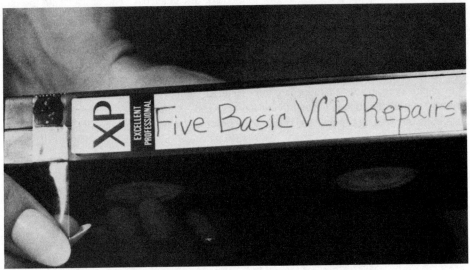

14-5   Placing a piece of tape over the missing tab.

15

Deactivating tape sensors

ALL VCRS HAVE TWO TAPE SENSORS, WHICH SERVE THREE PURPOSES. FIRST, the sensors protect the video tape from breaking when it reaches the end of the reel. Second, they signal to the unit that the video cassette is either at the beginning or end of the tape. Finally, the sensors stop and/or rewind the video tape.

It's necessary to have the unit running without a video tape inserted to perform certain procedures, so the tape sensors must be deactivated. Outside room light will penetrate into the sensors and stop the unit from playing. The light also can cause the unit to go automatically into fast forward or rewind, which can be disconcerting if you're adjusting or repairing the unit. In some cases, the VCR won't run at all because too much room light is reaching the sensors.

Tape sensor functions

There's a cassette lamp, a small light bulb or LED, located on the transport in the middle of the cassette carriage, as shown in Fig. 15-1. This lamp travels up inside the video cassette when cassette is inserted. On each end of a video tape there's transparent leader tape. When the video tape reaches the end, the light shines through the leader tape and through a small hole located on either end of the video cassette. The light then continues to travel to a tape sensor, which signals the unit to stop.

Topload VCRs

In most older models of VCRs, the video cassettes are loaded into the top of the unit. The tape sensors in topload VCRs are easily accessible. Follow the instructions in the following sections to locate and deactivate the tape sensors.

15-1 A cassette lamp.

Locating the tape sensors

The two pick-up tape sensors are located in a direct line from one another. There's one on each side of the cassette carriage. To locate the sensors, insert a video cassette. As the cassette holder drops down into its locked position, the front door on the video cassette opens up. Look along the top edge of this door (Fig. 15-2). Follow this line to each end of the video cassette until you reach the cassette carriage. Then look straight down, and you'll find a tape sensor.

Identifying a tape sensor

There are three types of tape sensors found in topload models. They will be a small black rectangular box with a hole in the center (Fig. 15-3), a round black rubber-type sensor (Fig. 15-4), or a long black plastic cylinder with a hole in the center (Fig. 15-5).

Deactivating the tape sensors

To deactivate the tape sensors, place a small piece of black electrical tape, over the hole on each sensor. Be sure the cassette holder is locked down without a video cassette being inserted. Now the unit will run without a video tape.

Frontload VCRs

The VCRs being manufactured today are all frontload models. In these models, the tape sensors are physically inaccessible; however, they still can be deactivated. The following sections will show you where the tape sensors are located in frontload VCRs and how you can deactivate the sensors.

15-2 Following the top edge of the cassette door to locate the tape sensors.

Locating the tape sensors

Locate the cassette holder in the carriage. On each side of the cassette holder is a small round hole, approximately ⅛ to ¼ inch in diameter (Fig. 15-6). When a video cassette is inserted and the cassette holder has dropped down into the carriage, these holes line up with the holes on the tape sensors. In some models, the tape sensors are mounted on the sides of the cassette holder and move with the cassette holder.

Deactivating the tape sensors

Place a ½ to ¾-inch piece of black electrical tape over each hole on the sides of the cassette holder, as shown in Fig. 15-7. You'll need to release the locking latches to get the cassette holder to drop down without inserting a video cassette. Proceed to chapter 17, "Making a frontload VCR run without a video cassette." After dropping the cassette holder down, the unit will run without a video cassette.

Replacing tape sensors

In order to replace the tape sensors, the cassette carriage must be removed. Refer to chapter 18, "Removing cassette carriages in front-load VCRs." On each end of the carriage is a small circuit board, as shown in Fig. 15-8. At either end of this board is a small clip. Pull the clip away and pull the board out. Locate the tape sensor (Fig. 15-9) on the component side of the board. Unsolder the two pins protruding through the board. (For detailed instructions,

15-3   First type of tape sensor.

15-4   Second type of tape sensor.

15-5 Third type of tape sensor.

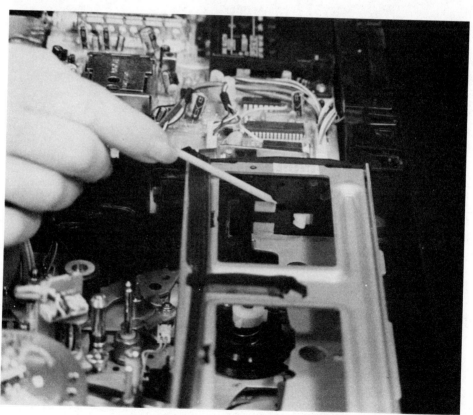

15-6 Tape sensor hole in the cassette holder.

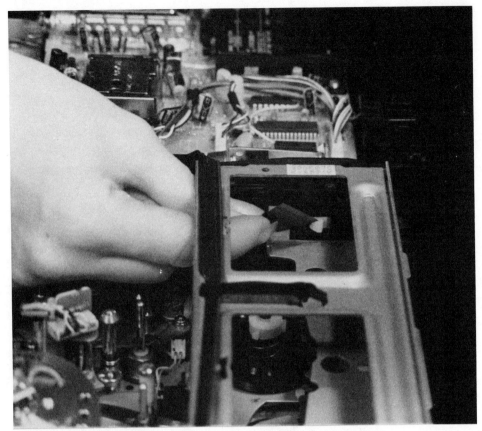

15-7   Placing black electrical tape over the tape sensor hole.

read the sections in chapter 24 on "Using a solder wick to remove old solder" and "Soldering pins to a circuit board.")

Each type of sensor has a little bubble on its side, as shown in Fig. 15-9. To insert the new sensor, the bubble must be pointing away from the circuit board. Make sure the bubble is facing the center of the hole in the cassette carriage. Remount the circuit board to the carriage.

Review

☐ There are three types of sensors in top-load VCR's.
- Small black rectangular plastic box.
- Round black rubber type.
- Long black plastic cylinder type.

☐ All sensors have a hole in the center.

☐ You can use the top edge of the video cassette door to help find the sensors in top-load units.

☐ Remember to cover both sensors with black electrical tape.

☐ In frontload units, cover the holes on each side of the cassette holder.

After completing all repairs, remember to take the black electrical tape off the tape sensors!

15-8 A tape sensor circuit board.

15-9 A tape sensor.

16

Removing
cassette carriages
in topload VCRs

OCCASIONALLY IT'S NECESSARY TO MAKE A UNIT RUN WITHOUT THE CASSETTE
carriage to perform needed repairs on the take-up, fast forward, and rewind systems or on
the carriage. These procedures are described in the following chapters. Having the unit running gives you better access to work on it and provides an important visual aide. Be sure to
read chapter 15, "Deactivating tape sensors," before removing the cassette carriage.

Removing and remounting the first type

In some cases, it's necessary to remove the cassette holder or the entire mechanism, called
the cassette carriage, to perform needed repairs. The first type of cassette carriage usually
is found in older units. The cassette carriage is mounted directly to the transport by four
mounting L-brackets, each with a Phillips head screw. One bracket is located in each corner
of the cassette carriage, as shown in Fig. 16-1. (I've shown you the location of one of these
mounting brackets and its screw.) Follow these next steps to remove the cassette carriage.
Push Eject to pop up the cassette holder and remove the four mounting screws. A magnetic
screwdriver comes in handy for retrieving the screws.

A counter belt usually runs over the top of the front right side mounting L-bracket, as
shown in Fig. 16-2. Using a long flathead screwdriver, pull this belt over to the side, to clear
the bracket. Now, pull the cassette carriage straight up and out. Figure 16-3 shows what a
removed cassette carriage looks like. You can see the four mounting L-brackets holding the
cassette carriage in place.

When a cassette holder is in a locked position, it activates a leaf type microswitch. This
microswitch signals the unit to run. Most microswitches are located on the front left side in
front of the supply spindle and are mounted to the transport, as shown in Fig. 16-4. Cut a
piece of black electrical tape approximately two inches long, push the arm on the switch
down, and place the tape over it. The tape will keep the arm down, as shown in Fig. 16-5.
Next, place a piece of black electrical tape over each tape sensor. Refer to the section in

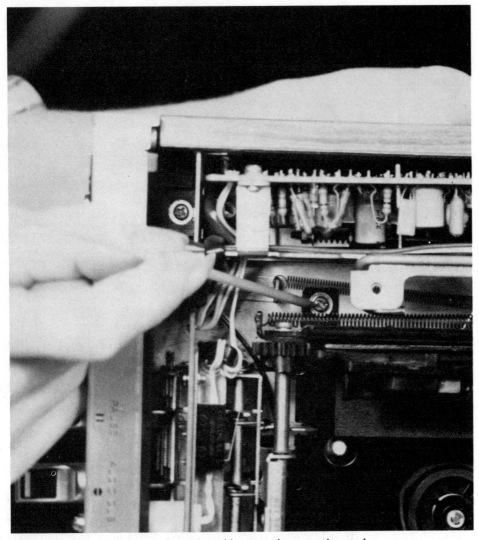

16-1 An L-bracket with a mounting screw inserted.

chapter 15 on "Topload VCRs." Now, the unit will run without a video cassette and cassette carriage.

When remounting this type of cassette carriage, make sure the ejection tab mounted on the front right side of the transport is pushed all the way back, as shown in Fig. 16-6. If you aren't sure you have the right piece, push Eject to make the tab move. Next, take the ejection bracket arm mounted on the bottom of the cassette carriage (Fig. 16-7) and push it forward, so the arm on the bracket goes in front of the tab on the transport. You need to do this simultaneously as you're remounting the cassette carriage to its original position. If you fail to do this, the eject mode won't work. Also, remember to move the counter belt over when you're remounting the carriage. If you fail to move the belt over, one of the mounting L-brackets will pinch the belt up against the transport and the tape counter won't work.

16-2   Locating the counter belt.

16-3   The location of the four L-brackets.

Next, align each mounting L-bracket with the screw holes on the transport. Then, replace and tighten down all screws.

The preceding type of models must be plugged in and turned on for the eject mode to function.

16-4　Locating a leaf switch.

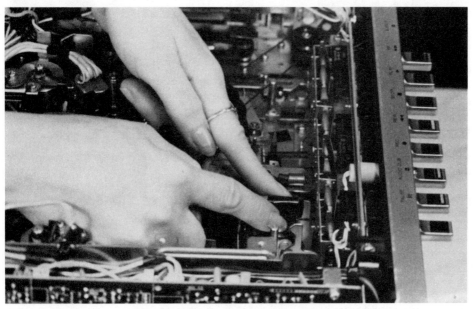

16-5　Taping down the arm on the leaf switch.

16-6 The ejection tab.

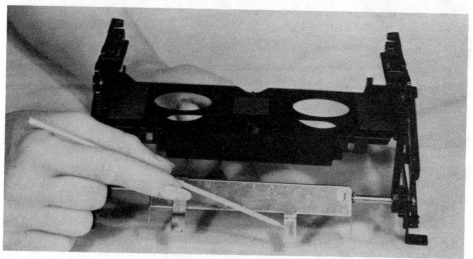

16-7 The ejection bracket arm.

Removing and remounting the second type

To remove this type of cassette carriage, it must be in the locked down position, so that you can unscrew the mounting screws. This type of carriage has four mounting screws, which are located through the holes in the bottom plate of the cassette holder, as shown in Fig. 16-8. After removing the screws, push Eject and the cassette holder will pop up. You might find a grounding wire mounted to the carriage by one of the mounting screws, as shown in Fig. 16-9. Simply move the wire to one side and pull the carriage up and out. This type of model is a direct drive unit and contains a microswitch. Over 90% of all VCRs have a microswitch. To make this type of unit run without a cassette carriage, cover both tape sensors with a piece of black electrical tape, as explained in the section on chapter 15 on "Deactivating the tape sensors" under the subhead "Topload VCRs."

To remount this cassette carriage, locate the two plastic pins coming up from the transport. There is one pin for each mounting bracket, as shown in Fig. 16-10. Notice that each mounting bracket for the carriage has three screw holes (Fig. 16-11). Match up the center hole in each mount to the pins coming up from the transport. While remounting the carriage, notice the ejection tab (Fig. 16-12), which is located on the bottom of the cassette carriage. Make sure this tab is in front of the leaf switch and behind the white ejection arm located on the transport, as shown in Fig. 16-13. Next, place the grounding wire over its proper screw hole. Push the cassette holder down into the locked position and replace the mounting screws. Recheck the grounding wire, making certain it's properly secured.

16-8   Four mounting screws under the cassette holder.

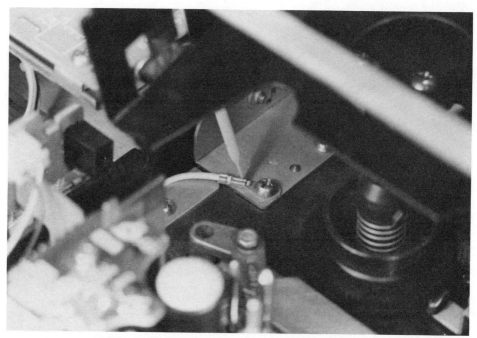

16-9 A grounding wire mounted to the carriage bracket.

16-10 Locating a plastic pin.

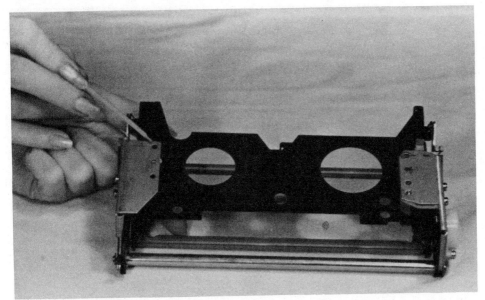

16-11 Locating the three screw holes on each bracket.

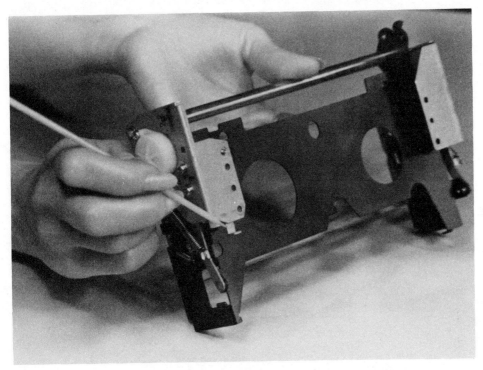

16-12 The ejection tab.

16-13 The correct positioning for the ejection arm.

Removing and remounting the third type

Push the Eject button to pop up the cassette holder. Remove the shield plate covering the video tape path. Refer to the section in chapter 3 on "Removing the shield plates covering the tape path." Part of the shield plate covers the carriage mounting bracket (Fig. 16-14). In this case, the plate covers only one bracket. In another model, the shield plate is all the way around the cassette carriage, covering both mounting brackets and screws. These shield plates are held on by either four or six mounting screws and are usually red in color. After removing the shield plate, you'll need to remove the two mounting bracket screws, as shown in Fig. 16-15. Look on the other side of the cassette carriage and you'll find another bracket like the one shown in Fig. 16-15. Remove these screws, then pull the cassette carriage straight up and out.

To make this type of model run without the cassette carriage, you must locate the hidden microswitch. To determine if you have this type of model, look at the front of the cassette carriage for a lock arm with a white glass roller around a pinlike shaft, as shown in Fig. 16-16. In some models, the pin might not have a glass roller. This pin points in either direction, depending on the type of model.

Position the cassette carriage over the top of the unit as you would to remount it. Look straight down from the glass roller and match it up with the transport. At that location, you'll find a white locking latch with a notch in it. Take your finger and push it down (Fig. 16-17) to activate the hidden microswitch and lock down the latch. Then, cover the light sensors with black electrical tape. Now, the unit will run without a video cassette and carriage.

There's another type of model that also has a lock arm with a pin on it but has no glass roller (Fig. 16-16). The pin is facing the opposite direction. The white locking latch on the

16-14   The shield plate covering the mounting bracket.

16-15   The two mounting screws for the cassette carriage.

16-16   A white glass roller around a shaft.

transport is usually located in the same place. Instead of swiveling, it pushes straight down into its locked position. Use the same procedure explained previously to locate the locking latch. Push straight down on the latch to lock and activate the hidden microswitch, as shown in Fig. 16-18.

To remount the carriage, locate the solenoid (Fig. 16-19). Push in with a small screwdriver on the shiny round shaft coming out of the end of the solenoid to release the locking latch to its normal position. Now, place the cassette carriage back into its transport and replace the mounting screws. Remove the black electrical tape from the tape sensors and replace the shield plate.

To release the locking latch in models with no solenoid, turn the power on and push Eject and the latch will pop back up. Then, replace the carriage. All locking latches must be in the up position to remount the carriage.

There are three ways these types of cassette carriages might be mounted to the unit. In the first, one side of the carriage is mounted near the top on a separate bracket, and the other side is directly mounted to the transport. Both mounting brackets are shown in Fig. 16-20. In the second way, both sides of the carriage are mounted near the top, like the bracket shown on the left side in Fig. 16-20. In the third way, both sides are directly mounted to the chassis, like the bracket shown on the right side in Fig. 16-20. No matter what type of unit you have, remove the four mounting screws and pull the carriage straight up and out.

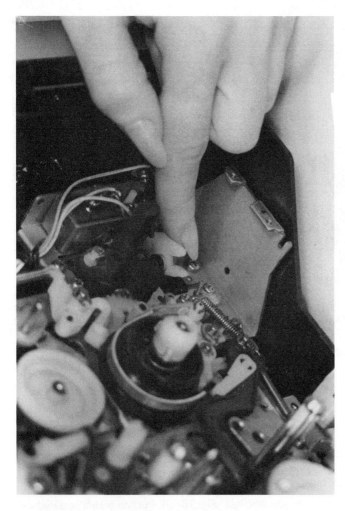

16-17 Pushing down the hidden microswitch.

Removing and remounting the fourth type

Push Eject to pop up the carriage. This model has six mounting screws holding the cassette carriage into place, as shown in Fig. 16-21. On each side, you'll see a brass grounding leaf spring attached to the center mounting screw. Remove the screw, detach the spring, and lift it off. Repeat this procedure on the other side. Remember how the springs were attached.

After removing all six mounting screws, look to the front left side of the carriage for a protruding pin going through a governor arm, as shown in Fig. 16-22. If there's a C-ring on the end of the pin, remove it. Refer to the section in chapter 7 on "Removing C-rings" under the subhead "Finding the lubricating points." Slide the carriage slightly to the right until the governor arm detaches from the pin. Then, lift up on the front of the carriage and slide it forward until it clears the lip on the mounting bracket at the rear of the unit. Now, lift the carriage straight up and off.

After removing the carriage, look for the microswitch. In this model, the location of the microswitch is the same as that shown in Fig. 16-4. Take a piece of black electrical tape and

16-18   Locking down the hidden microswitch.

16-19   Locating the solenoid and the protruding shaft.

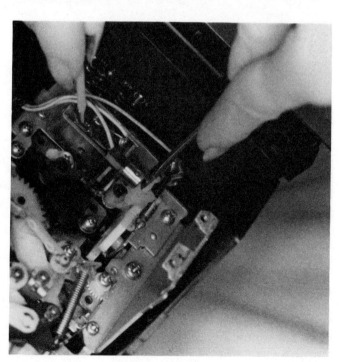

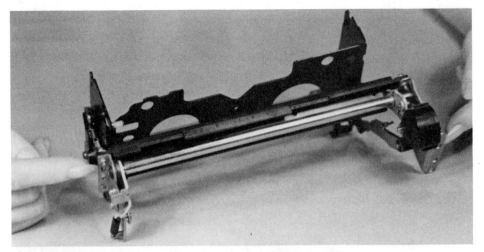

16-20 The mounting brackets.

16-21 Locating the six mounting screws.

tape down the arm on the microswitch, as previously shown. Place black electrical tape over both sensors. Now, the unit is ready to run without a cassette carriage.

To remount the carriage, slide the carriage mounting brackets back into their original position at the rear of the unit. Then, slide the front of the carriage slightly to the right and place the governor arm back onto its pin. Slide the carriage slightly back to the left. Now, line up the holes on the carriage mounting brackets near the back of the unit with the mounting screw holes on the chassis. Replace the mounting screws and brass springs.

16-22   The pin protruding through the governor arm.

Levers and latches

Although all carriages are mounted the same, they might have different types of levers, and latches similar in principle. The first step is to remove the shield plate covering the tape path. Push Eject, then look under and around the carriage for all levers and latches attached to it. You might want to push Eject and then push the cassette holder back down into its lock position several times, so you're sure you've found all the attached levers and latches.

Find the mounting screws and remove them. Pull the cassette holder straight up and out. If it doesn't come out easily, look to see what's holding it back and carefully disconnect it.

Some models have a large cassette lid attached to the top of the cassette holder that is hard to see around it. If this is the case, remove the cassette lid. Refer to the section in chapter 9 on "Removing a cassette lid" under the subhead "A jammed video cassette in a topload VCR."

Review

☐ Plug in the unit and turn it on.

☐ Look at the latches and all levers that move when you push the Eject button.

☐ Remove the mounting screws.

☐ Pull the cassette carriage straight out watching for and remembering all the parts that are attached to the cassette carriage, so you can reconnect them later.

☐ Cover the tape sensors.

☐ Tape down and activate the microswitches, if needed.

☐ When replacing the carriage, replace all tabs and pins into their proper positions before replacing the mounting screws.

17

Making a frontload VCR run without a video cassette

TO PERFORM THE PROCEDURES EXPLAINED IN THE FOLLOWING CHAPTERS, IT will be necessary to be able to place a unit into play, fast forward, or rewind without a video cassette inserted. The locking system must be deactivated to make a unit run without a video cassette inserted. There are three different types of locking systems. Find the locking system that matches your unit.

The light sensors in the first model are located on each side of the cassette holder. Place a small piece of black electrical tape over them. Refer to the section in chapter 15 on "Locating the Tape Sensors" under the subhead "Frontload VCRs." Push Power.

Disengage the locking latches on the cassette carriage to drop the cassette holder down. The cassette holder is locked down on both sides by these latches. The latches have two locking positions, which are the two square holes on each side of the cassette carriage. The latches are spring-loaded and will pop up into the holes, which keep the cassette holder from moving (Fig. 17-1).

There are two locking levers that release the latches. The levers are located at the rear of the cassette holder (Fig. 17-2). The other locking lever is on the opposite side of the cassette holder. When a video cassette is inserted, it pushes against these levers. When the locking levers are pushed forward, the locking latches drop down out of each hole on the carriage, unlocking the cassette holder.

To drop the cassette holder down, without a video cassette, take a 1-inch piece of black electrical tape and securely tape each lever back, as shown in Fig. 17-3. Then, gently push back on the cassette holder, reaching through the front door, as shown in Fig. 17-4. When you feel the cassette holder move on its own, you'll hear the ejection motor start up and the cassette holder will drop down automatically. Now, the unit will run without a video cassette.

In the second type of model, the locking latches and levers are located in the same places as in the first type. Refer to Figs. 17-1 and 17-2. The difference is that the two levers are in the back edge of the cassette holder, so there's nothing to tape the levers to.

Place black electrical tape over the tape sensor holes, as previously shown. Reach down

17-1 The locking latch protruding through the hole in the carriage.

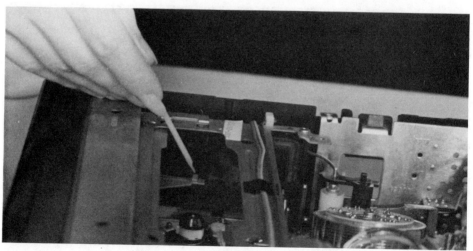

17-2 The locking lever.

through the top of the cassette carriage and push both locking levers back simultaneously, as shown in Fig. 17-5. While pushing on the locking levers, gently push the cassette holder back until you feel the unit start moving under its own power. As soon as you feel the unit start moving under its own power, pull your fingers out. The cassette holder might grab your fingers if you wait too long to let go. You might want to use two pencils, instead of your

17-3   Taping back the locking lever.

17-4   Pushing back the cassette holder.

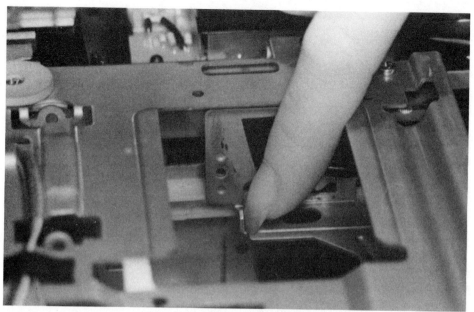

17-5 Pushing back both locking levers.

17-6 The leaf switch on top of the cassette carriage.

fingers, to push on the locking levers. Then, the cassette holder automatically drops down into its loaded position. The unit is now ready to run without a video cassette.

The third model has no locking latches. Instead, it uses a leaf switch, which is located in the center of the top bracket of the cassette carriage (Fig. 17-6).

17-7   Taping down the leaf switch.

Place black electrical tape over each tape sensor hole. Reach under the top bracket and place a piece of black electrical tape over the leaf part of the switch. Now, push up on the leaf part of the switch to activate it, as shown in Fig. 17-7. Secure both ends of tape by folding them over the top of the bracket. Turn the power on and the cassette holder automatically drops down into its loaded position. Now, this type of model is ready to run without a video cassette.

In other models, you'll find two leaf switches located on top of the cassette carriage, one on each side. To activate this model, tape down both of these leaf switches. Turn the power on and the cassette holder will automatically drop down into its loaded position.

When you've completed the repairs and want to bring the cassette holder back up, re-move the tape from the leaf switches and push Eject. Remember to remove the tape from the light sensor holes on the inside of the cassette holder.

Review

☐ Place tape over each light sensor hole.

☐ Look for the latches on top of the cassette carriage.

☐ Look for the locking levers on the cassette holder that release the latches.

☐ Tape back the locking levers.

☐ Check for a leaf switch under the top cassette bracket.

☐ Tape down the leaf switch to activate it.

☐ Turn the power on.

☐ Push the cassette holder back until it automatically drops.

18

Removing cassette carriages in frontload units

IT'S NECESSARY TO REMOVE THE CASSETTE CARRIAGE TO REPAIR A GEAR broken on the carriage, A bad tape sensor, or the housing loading motor. In some models, it's necessary to remove the carriage to reach the idler wheel and clutch assembly or to replace a tire on a wheel. It also is helpful when unjamming a jammed video cassette.

Preparation for removing cassette carriages

To remove the cassette carriage, you first need to remove the front cover. Refer to the first five sections of chapter 35. In 50% of all VCRs, you'll find a bracket across the top front of the carriage, mounted to the chassis, as shown in Fig. 18–1. There might be one screw on each end of the bracket. Remove the screws. On the right side of the bracket is a black clip (Fig. 18–1). Push the clip in to release the bracket, then lift this side of the bracket straight up until you clear the front cover latch on the right side of the bracket. Push the bracket to the left, to release the lip going into the chassis on the left side of the bracket. To remount the bracket, reverse this process.

There are two more ways the screws might be mounted to the bracket. The screws will be in the same location, but the carriage also will be mounted with these screws. Secondly, the screws will be mounted through the top of the front cover and through the bracket. When you've removed the front cover, the bracket will appear to have no screws. In one model, the bracket completely crosses the unit, containing three screws. Remove the screws and pick it straight up.

Most cassette carriages have a white, green, red, brown, or black plug with eight to ten wires protruding from it. This plug needs to be unplugged. Usually the plug is located on the back right side of the carriage, on a small circuit board, as shown in Fig. 18-2. In other models, the plug is on the left side at the back of the carriage. In some models, you might find two plugs on the right side, instead of one. One will be large in size, and the other small. Unplug both of these plugs.

18-1   Locating the bracket across the front of the carriage.

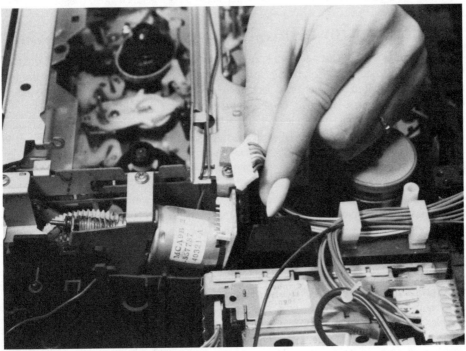

18-2   Locating the plug on a small circuit board.

Another system used has a ribbon attached to the carriage, instead of wires and a plug. Although the ribbon does disconnect, it isn't necessary to uncouple it. (The ribbon is difficult to reconnect.) After removing the carriage, lift the carriage with the ribbon intact, flip it to the right, and lay it on top of the unit. Be sure to place a towel between the carriage and the top of the unit, for insulation.

Cassette carriage removal

There are six different procedures for removing the cassette carriage in the various types of VCRs. Read each of the following sections and follow the instructions that relate to the model that you have.

First type

This procedure is the most common one used to remove a cassette carriage. You'll see a white plug at the back right corner of the carriage. Unplug this plug. At the back of the carriage, there are two screws mounted to the transport, one on each side (Fig. 18-3). The screws are usually red in color, but are sometimes gold. Remove the screws. Pull the carriage back about ½ inch to release the front latches. Pick the carriage straight up and out.

To remount the carriage, align the two pins on the bottom front portion of the carriage with the slots in the transport. Figure 18-4 is an illustration of a pin. Slide the carriage forward until the screw holes on the carriage align with the screw holes on the transport. Replace the two screws and plug in the plug.

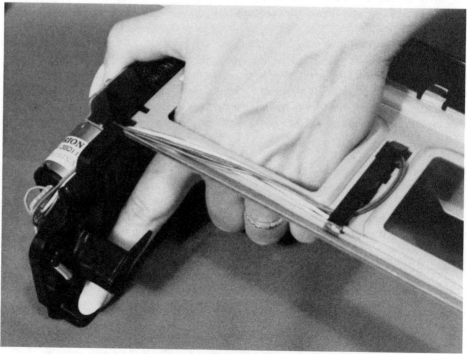

18-3   Locating the mounting screws into the transport.

18-4 Locating the pins on the front of the carriage.

Second type

This model has a bracket on each side at the base of the carriage, with three screws on each side, as shown in Fig. 18-5. These brackets mount the carriage to the transport. If the unit has three large screwheads, remove only the middle screw on each side. If the unit has two small screwheads and a large screwhead in the middle, then remove all three screws. Lift the carriage straight up and out, then unplug the plug found on the right side of the carriage. To remount the carriage, reverse this process.

Third type

The third type of model is similar to the first type, except that the third type has six mounting screws instead of two. Unplug the plug. Remove the four top screws mounted to the chassis on each side of the carriage, as shown in Fig. 18-6. Remove the two screws at the back of the carriage mounted to the transport. Slide the carriage a ½ inch back to release the latches. Pick the carriage straight up and out. To remount reverse this process.

Fourth type

Unplug the plug. This type of model has six mounting screws. Two of the screws protrude through the bracket across the top front of the carriage, as explained in the first section of this chapter. Two other screws are mounted about half way back, on top of the carriage, one on each side. These two screws are mounted into the chassis. The last two screws are mounted to the transport, as explained in the section on removing the first type of cassette carriage. Remove all six screws. Remove the front bracket and lift the carriage straight up and out. To remount the carriage, reverse this process.

Fifth type

Unplug the plug. This type of carriage is mounted with four screws. The front two screws are mounted through the bottom of the carriage into the transport. You'll need to slide the cassette holder back about ½ inch to remove these screws. The back two screws also are mounted to the transport, as explained in the section on removing the first type of cassette

18-5 Locating the three mounting screws on each side of the carriage.

18-6 Locating the four mounting screws on top of the carriage.

carriage. Remove the four screws and lift the carriage straight up and out. To remount the carriage, reverse this process.

Sixth type

Unplug the plug. This type of carriage is mounted with four mounting screws. All four screws are mounted on top of the carriage and into the chassis. There is one screw in each corner, as shown in Fig. 18-6. Remove the four screws and lift the carriage straight up and out. To remount this type of carriage, reverse this process.

Review

☐ Remove the front cover.

☐ Remove the bracket across the top front of the carriage.

☐ Remove the plug located at the back of the carriage.

☐ Remove the mounting screws:
  - Located at the back of the carriage, mounted to the transport.
  - Mounted to the bracket on top, in front of the carriage.
  - Located on top of the carriage, one on each side or two on each side.
  - Brackets mounting the base of the carriage to the transport on each side of the carriage.
  - In the front bottom of the carriage, mounted to the transport. You'll need to push the cassette holder back to remove the screws.

☐ In most models, push the carriage back about ½ inch to release the latches. Then remove the carriage.

☐ In all other models, the carriage lifts straight out.

19
Setting the torque
on a video tape

THE BACK TENSION GUIDE KEEPS THE CORRECT AMOUNT OF TORQUE ON THE video tape while it is playing. The back tension guide is located on the video tape path. Torque is the amount of pressure being applied. The back tension guide pole has the appearance of an upside down pencil (Fig. 19-1). It's mounted on a moving bracket. There are two ways the torque can be off. First, the bracket can become bent and hang up, causing no torque on the video tape. The second way is that the spring attached to the tension guide bracket can become stretched out after a period of time, causing too little torque on the video tape.

Detecting a back tension guide problem

You have to determine whether you have a back tension guide problem, a dirty video head, or an A/C head that's out of alignment. These three problems have some common symptoms. To find out, connect the TV monitor to the VCR and insert a rented video tape. Push Play and watch and listen to the TV monitor. If you have a dirty video head, you'll have a partially distorted or completely snowy picture. Refer to the section in chapter 5 on "Dirty video heads." If the A/C head is out of alignment, you'll have a belt of lines, starting at the bottom of the picture and moving up the screen. Proceed to the section in chapter 14 on "Lines floating through the picture" under the subhead "Video."

There are three ways to detect if the tension guide is hung up or out of adjustment. First, the TV monitor shows a completely washed out, distorted picture, or you have many lines and dashes across the picture. The audio portion has a buzz or distortion in it. Second, the picture is stable with a normal clear picture, then becomes unstabilized with a distorted picture, but returns to a clear picture only to distort again. Each cycle takes between 5 and 15 seconds. The audio portion also will sound like it's speeding up and slowing down, or the audio might have a distortion in it that clears up along with the picture. Third, you see a configuration of white rectangular and square boxes, resembling a large city. The configura-

19-1   The back tension guide pole.

tion is a long and narrow bar across the screen that starts at the top of the screen, works its way down, then disappears. It'll keep repeating this process. It might also remain stationary at the top of the screen, appearing and disappearing.

If you see and hear these types of symptoms, check to see if the tension guide pole is hung up or tilted. As you follow the following steps, have the unit in play with a video cassette inserted.

Checking the tension guide pole

Compare your unit to Fig. 19-1. While the unit is in play, see if the center of the tension guide pole is approximately 1 to 5 mm left of the center of the supply tape guide. The supply tape guide is directly behind the guide pole. Tape guides vary in width and looks like an empty spool of thread. The tape guide pole should be perfectly straight up and down, not tilted. If the pole is not tilted, place your finger on top of it while the unit is running, as shown in Fig. 19-2. Move the tension guide pole approximately ½ inch to the left, as you face the front of the unit. Check for freedom of movement by releasing the tension guide pole to see if it pops back to its normal position. Now push the tension guide pole to the right for an inch or so to check for freedom of movement in this direction. Again, release the tension guide pole to see if it pops back into its normal position. If it's not bent, the tension guide pole will pop back. By following this procedure you'll learn whether the tension guide bracket is bent, hung up, or tilted.

If you find the tension guide pole isn't hung up or tilted, follow the steps in the following sections.

Using a TV monitor
to check the tension guide pole

Be sure the TV monitor is connected properly to the VCR and push Play. Place your finger onto the tension guide pole as previously shown. Start by slowly pushing the guide pole to

19-2   Placing a finger on top of the tension guide pole.

the left, periodically stopping approximately every ⅛ inch and holding that position for 20 to 30 seconds. Each time you stop, check to see if the picture clears up on the TV monitor. Continue this procedure until you can go no further. If the picture doesn't clear up, release the guide pole and let it pop back to its normal position. Then slowly push to the right, again periodically stopping to check the picture. Usually you'll find a spot to the left of the normal position that clears up the picture.

It's important to remember when using this method that the picture must completely clear up, or stop pulsating, while holding the tension guide pole in one of these positions. If it does, proceed to the section in this chapter on "Adjusting the back tension guide." On the other hand, if you can't stop the pulsating, refer to the section in chapter 14 on "Lines floating through the picture" under the subhead "Video." If you can't clear up the whole picture and stationary lines are running horizontally at the top or bottom of the picture, then your roller guides are probably out of alignment. Do not make this adjustment. Do a video tape path alignment instead. Proceed to chapter 20, "Video tape path alignment."

Repairing a bent tension guide pole bracket

If the tension guide pole bracket is bent, it'll drag on the top of the transport or on some other bracket near the tip of the bracket where the guide pole is mounted. The place where the bracket usually gets bent or twisted is pointed out in Fig. 19-3. In other cases, the bracket could be twisted up and not be dragging on anything but will cause the guide pole to be tilted. This misalignment causes the wrong amount of torque on the video tape. You'll need to remove the bracket to straighten it out. Follow these next steps.

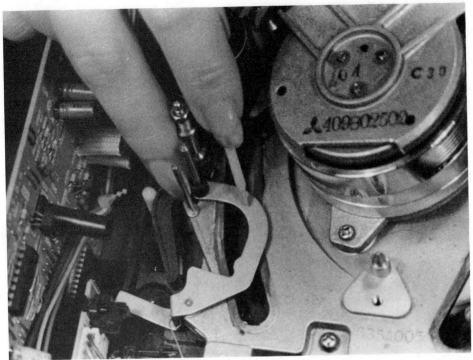

19-3 The point the guide pole bracket gets bent or twisted.

First, remove the cassette carriage. Refer to chapter 16, "Removing cassette carriages in topload VCRs," or chapter 18, "Removing cassette carriages in frontload VCRs." Next, remove the C- or E-ring on the shaft holding the bracket on (Fig. 19-4). Refer to the sections in chapter 7 on "Removing E-rings" and "Removing C-rings" under the subhead "Finding the lubricating points." Then, pull the bracket straight up and off. Next, you'll need to remove the tension guide spring. Holding the bracket in your hand, slip the hook on the end of the spring off the notch on the bracket, as shown in Fig. 19-5. You'll find a plastic arm with a metal band attached to it. This band wraps around the supply spindle and is called the break shoe band.

On some models there is no C- or E-ring holding the bracket in place. What you'll find is a white plastic mount. Pull straight up on the mount for removal. This type of mount snaps on and off.

When the unit is in play, this band puts pressure on the supply spindle and causes drag. The drag causes torque on the video tape. The band has to be disconnected to remove the tension guide bracket. There are two ways to remove the band. One type of unit has a small Phillips head screw located on top of the break shoe arm. By removing this screw, the band will lift right off. Figure 19-6 shows where this screw is located. The second type of unit has a C- or E-ring located on the back side of the tension guide bracket, as shown in Fig. 19-7. Just remove the C- or E-ring and the band will come off. At the opposite end of the band, you'll find another adjustment screw. You do not need to remove this screw to remove the band. This adjustment screw is very critical. You could mess up the torque adjustment if you turn that screw.

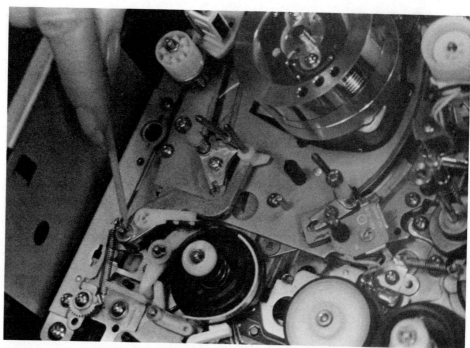

19-4   The C-ring holding the back tension guide bracket.

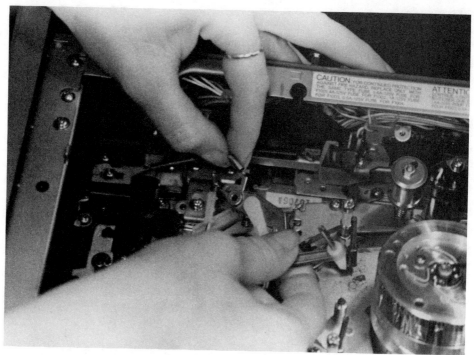

19-5   Removing the spring.

19-6 Where the screw would be located on the break shoe arm.

19-7 The C-ring located on the backside of the tension guide bracket.

After removing the tension guide bracket, you can either straighten the bracket out or take it to a local electronics supply center, along with the make and model number of the VCR, for a replacement. To remount the bracket, simply reverse the procedure.

Adjusting the back tension guide

The majority of all models have a spring connected to the tension guide pole. If you had to move the tension guide pole to the left to clear the picture, then you'll need to tighten the spring. On the other hand, if you had to move the guide pole to the right, then you'll need to loosen the spring. You'll find the adjustment at the base of the spring. Read the following types of adjustments and find the one pertaining to your unit.

The first type of adjustment screw has a white plastic bracket with notches on it and a screw in the middle of it, as shown in Fig. 19-8. (I've removed the cassette carriage for a better view. It's not necessary to remove the carriage, to make these adjustments.) To make the adjustment, eject the video cassette and loosen the screw on the bracket. Then, move the white plastic bracket one or two notches toward or away from the spring. Usually you'll be tightening the spring. Retighten the screw, insert a video cassette, and press Play. Now look at the TV monitor to see if you have a clear picture. If not, go back and readjust the spring tighter, a notch or two at a time.

The torque on the video tape is important. Just make little adjustments, one at a time, until the picture clears up. Do not over tighten the spring.

There are two other ways to make the adjustment. The springs in some models are attached to a metal bracket, which can be slid back and forth by loosening the holding screw (Fig. 19-9). In other models, the bracket can be turned to tighten the spring by loosing the holding screw, as shown in Fig. 19-10. In these models adjust the bracket while the unit is in play.

19-8   First type of adjustment.

19-9   Second type of adjustment.

First, loosen the holding screw and push Play. If you moved the tension guide pole to the left, move the bracket away from the spring a notch at a time. Stop for a minute or so, to see if the picture stabilizes or clears up. If not move it another notch and stop, repeating this procedure until the picture clears up. Do not over tighten the spring.

There's one more different type of model. In this type of model, the spring is attached directly to the notches on the transport (Fig. 19-11). Use the same procedure to adjust the spring, but this time move the spring itself from notch to notch. Remember to check the picture each time you move the spring, until the picture clears up.

In some of the older models, you can clear the picture by moving the guide pole over. When you go to make the adjustment, the spring is tightened all the way. In this case, remove the spring and take it to a local electronics supply center for a replacement. After replacing the spring, set the adjustment all the way toward the spring. By doing this, you are relaxing the spring. Now, follow the above procedure, tightening the spring notch by notch until the picture clears up.

The models without a spring attached to the tension guide bracket will have to be adjusted by adjusting the brake shoe band.

The brake shoe adjustment

You'll find a metal band wrapped around the supply spindle with felt inside it. This is the brake shoe band. On the end of the band, you'll find a plastic or metal bracket with notches and a holding screw, as shown in Fig. 19-12. When the felt has worn down and the spring has been replaced, you'll need to make this adjustment. You also will need to make this adjustment if the unit has no spring.

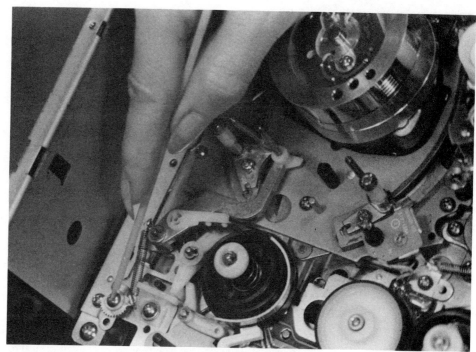

19-10 Third type of adjustment.

19-11 Fourth type of adjustment.

19-12 The brake shoe band adjustment.

The majority of the time, you don't have to make this adjustment. If you have tried to make the proper adjustment but the only way to clear the picture is by applying pressure to the guide pole, then you'll need to do the brake shoe band adjustment.

First, move the spring to its center position. Loosen the brake shoe band screw, move the adjustment one notch to the left, and retighten the screw. Insert a video cassette and push Play. See if the picture has cleared up. If not, then move the band adjustment one more notch over to the left and check the picture again. Overtightening causes the tape to slow down or stop.

There's a tool that is helpful, but it's quite an investment. This tool is called a Tentelo Meter, which measures the grams of pressure on a video tape. The meter's owner's manual tells you the amount of grams to set on the video tape.

Review

☐ First determine if the problem is a dirty video head, misaligned roller guides, a misaligned A/C head, alignment, or a misaligned back tension guide.

☐ To determine which problem it is, place your finger on top of the tension guide pole and move it to the left and to the right slowly, periodically stopping to check if the picture clears up.

☐ If the picture clears up, the back tension guide is out of adjustment. If you can't remove the stationary lines, the roller guides are out of alignment. If the picture stays completely distorted or snowy, then you have a dirty video head. If the picture continues pulsating, then the A/C head is out of alignment.

☐ When tightening up the adjustment spring, move it notch by notch until the picture clears up.

☐ Do not over tighten the spring.

☐ Make sure the back tension guide moves freely before making this adjustment.

20

Video tape
path alignment

VIDEO TAPE PATH ALIGNMENT IS THE MOST COMMON REPAIR PROBLEM. I highly recommend you read this chapter carefully before performing this adjustment.

Detecting misaligned roller guides

Connect the TV monitor to the VCR and turn it on. Insert a rented video cassette. Don't use a video cassette you've recorded. Push Play. You'll notice a horizontal belt of lines across the screen. This belt of lines is in either the upper or lower portion of the TV screen, as shown in Fig. 20-1. You might have two or more belts of lines across the TV screen (Fig. 20-2). You might find that ⅔ of the picture is washed out and distorted, in either the upper or lower portion. Locate the tracking adjustment on the face panel of the VCR. The tracking adjustment is used to remove the lines from the picture. When you try to remove the lines, you'll notice they'll either move up or down the screen, but still remain.

Another problem you might run into occurs when someone lends you a video cassette they've recorded and lines appear upon playback in your unit. To determine if your machine is out of alignment or if theirs is, place the video cassette in a third unit. If it has lines in playback, then the bad machine is from the person you borrowed the video cassette from. If the third machine doesn't have lines, then the bad machine is yours.

Setting the tracking adjustment

To properly perform the necessary adjustment, the tracking control must be centered. On newer models, the adjustment is either a dial or a slide control. As you move the control, you'll feel an indentation or niche, which is the center position. Set the control to the center position. In older models, the tracking adjustment is a knob with a click position. Rotate the knob until you hear the click and leave it there in the center position.

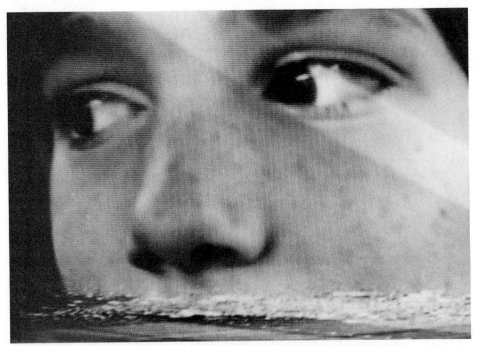

20-1   A belt of lines across the bottom portion of the screen.

20-2   Two belts of lines across the bottom portion of the screen.

The terms "load" and "unload"

A tape is unloaded if it has been inserted into the unit but and the video tape is still in the video cassette. A tape is loaded if you have pushed Play and the unit has pulled the video tape out of the video cassette and onto the video tape path. Keep these terms in mind, you'll be reading about them frequently.

Identifying roller guides and V-mounts

The supply roller guides are on the left hand side of the video drum as you face the front of the unit. The other roller guide is the take-up roller guide, which is on the right side of the video drum (Fig. 20-3).

Each roller guide is adjustable by turning the post. The post is the part of the guide with the white glass roller on it. The glass roller is the part that the video tape comes into contact with. In order to make these adjustments, you'll need either a wide thin-tipped flathead screwdriver or a 1.5mm, 1.27mm, or 0.89mm allen wrench. All roller guide posts have a fitting on the top, which the screwdriver or allen wrench fits into.

The first purpose of the roller guides is to load the video tape onto the video tape path. The second purpose of the roller guides is to keep the video tape aligned to the video drum, so that the video heads pick up the correct image.

Load the unit and push the Play button, so you can see the roller guides in their locked positions. Notice the roller guides are locked into position by V-mounts, one on each side of

20-3   The supply roller guide (left) and the take-up roller guide (right).

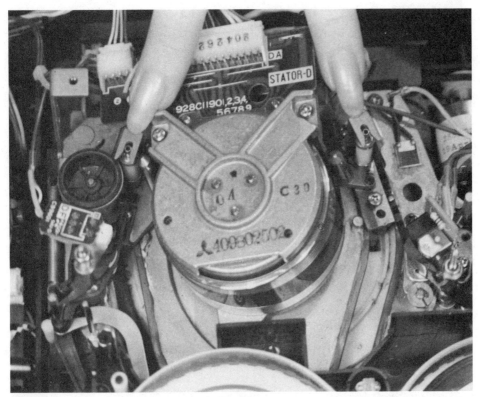

20-4 The V-mounts.

the video drum, as shown in Fig. 20-4. In order to make these adjustments, the roller guides must be in the locked V-mount position.

If you insert a video cassette and the unit goes into fast forward or rewind instead of play, you'll need to prepare the unit to make this adjustment. To prepare the unit, refer to the section in chapter 15 on "Deactivating the tape sensors" under the subheads "Topload VCRs" or "Frontload VCRs." You want to deactivate the sensors so the unit won't stop or unload while you're trying to align the tape path.

Loosening the base lock nuts

All roller guides are locked down by a lock nut that has to be loosened to adjust the roller guides. There are two methods used to lock down the roller guides. The first type is held in position by an allen nut, located on the back side at the base of the roller guide, as shown in Fig. 20-5. This nut has to be loosened before the roller guides are loaded on their V-mounts. Use a 1.5mm allen wrench to turn the lock nut. Turn it counterclockwise approximately ½ turn.

The lock nut might be found in a different location, protruding out the back of the brass base of the roller guide post, as shown in Fig. 20-6. You need to use a 1.27mm or 0.89mm allen wrench or a small-tipped Phillips head screwdriver to loosen this nut. It has to be loosened before the roller guides are loaded on their V-mounts. Turn the nut counterclockwise

20-5 First location for a lock nut.

20-6 Second location for a lock nut.

½ turn to loosen the roller guides. If the lock nut is up against the base of the roller guide, take hold of the roller guide and pull up on the post. If the roller guide lifts up, then the post is broken. Proceed to the section "Broken roller guide post."

The other way the roller guides are locked down is as shown in Fig. 20-7. The lock nut is located on top of the roller guide base, directly in front of the roller guide post. You'll find a Phillips head screw directly in front of the lock nut. *Do not touch it.* Insert a video cassette and push Play. This type of model has to be in the loaded position in order for you to be able to turn the lock nuts. Insert a 1.5mm allen wrench and loosen the base lock nut approximately ½ turn counterclockwise while the unit is running.

All models have a tension roller partially covering the base lock nut. To reach the base lock nut, gently pull the tension roller to one side, insert the allen wrench, and loosen the nut. (See Fig. 20-8) In some models, there also is a tension roller on the take-up side, and the same procedure must be followed.

In one model the base lock nut is in front of the roller guide. The nut is horizontally mounted into the base of the roller guide. A 1.27mm allen wrench is required to adjust it. This type of base lock nut must be loosened before loading the video tape onto the V-mount.

Broken roller guide post

Roller guide posts only break in models with a brass base roller guide post, as shown in Fig. 20-6. The brass base is compressed into the roller guide mount, the bonding breaks, and the post slips down.

20-7 Third location for a lock nut.

20-8   Pulling the tension roller to one side to insert the allen wrench.

Pull up on the post. Look for a fine line caused by tarnish to find the post's original position. You also could look at the other roller guide and position it to the same height. The top of the roller guides should be at the same height. Eyeballing this adjustment is fine; it doesn't have to be exact. Make sure the lock nut is pointing straight back, approximately 1mm above the roller guide mount. Once it is set, don't touch it. Take a tube of super glue and run a bead around the base of the roller guide post. (Avoid placing any glue on the locknut.) Let dry until it is set.

Adjusting the base lock nut

Regardless of the type of lock down system found in your unit, follow these next steps. Place either an allen wrench or flathead screwdriver on top of the roller guide post. You should be able to turn each roller guide back and forth smoothly. If it feels a little sticky, go back and loosen the base lock nut ¼ turn. Recheck the smoothness of movement of each roller guide post. It's important that the roller guides turn smoothly.

After checking the roller guide posts for smoothness of movement, leave the unit in play. Look carefully at the top of each roller guide post. The torque on the video tape might cause the guide posts to spin or turn. If either one does, tighten the base lock nut slightly. This adjustment is extremely temperamental. If you make the adjustment too loose, the roller guide you just aligned will go out of alignment after you remove the tool.

There's one more critical point, as shown in Fig. 20-9. Units with a top base lock nut also have an angle nut adjustment behind the V-mount (Fig. 20-9). Do not confuse this nut with the lock nut. The angle nut is factory-set. *Under no circumstances should it be turned.* If

20-9 The angle nut adjustment.

you've already turned this adjustment, the roller guide alignment won't work. Proceed to chapter 21, "The angle adjustment."

Roller guide functions

The roller guides pull the video tape out of the video cassette onto the video tape path. When the unit is in play, the supply roller guide aligns the video tape to the left side of the video drum. The take-up roller guide aligns the video tape to the right side of the video drum. Both roller guides align the video tape to the video heads.

Adjusting the roller guides

On the lower video drum, you'll find the tape edge guideline (Fig. 20-10). Notice that the slightly protruding edge runs on a downward angle from left to right. Adjust the roller guides, so the bottom edge of the video tape sits on the tape edge guideline.

Push Play to load the video tape. Look down at the lower video drum. Look for a slightly protruding shiny edge at the bottom of the video tape, as shown in Fig. 20-11. The shiny protruding edge should be visible all the way between the two roller guides. The bottom of the video tape should just touch the top of the tape edge guideline. If you don't see this edge all the way around (Fig. 20-12), locate the roller guide post that is closest to where the video tape covers the edge and slowly turn it counterclockwise. This procedure will move the video tape further up the video drum. When the shiny edge is exposed, stop turning the adjustment.

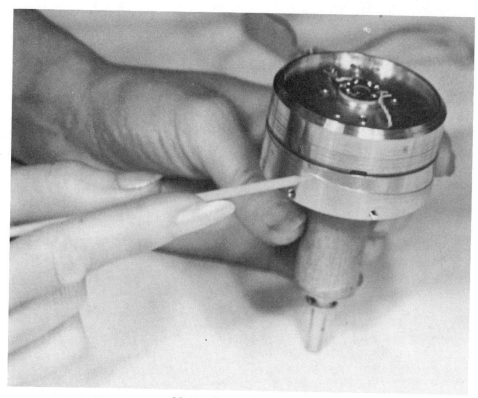

20-10 Tape edge guideline.

20-11 The protruding edge at the bottom of the video tape.

Now, align each roller guide to the tape edge guideline. Start with the take-up roller guide post, the one on the right side of the video drum. Insert the proper tool and turn it clockwise. This adjustment moves the video tape down. As the tape moves down, keep an eye on the shiny protruding edge and on the bottom edge of the video tape. As you slowly turn the roller guide clockwise, at some point, you'll see the video tape wrinkle. These wrinkles flow with the video tape around the video drum and toward the take-up roller guide, starting from a point 1 to 1½ inches away from it (Fig. 20-13). If you turn the adjustment too fast, the video tape will pop over the shiny protruding edge instead of wrinkling.

20-12   The shiny protruding edge not visible.

20-13   Wrinkling of the video tape at the protruding edge.

When you start to see the wrinkle, stop turning the adjustment. Now, turn the adjustment counterclockwise until the wrinkle just disappears. Turn it in the same direction for another ¼ turn and stop. The guide now is properly set.

Follow the same procedure with the supply roller guide, the one on the left side of the video drum. The only difference you'll find is that the wrinkles at the bottom of the video tape move away from the roller guide, rather than toward it.

You can save some time when adjusting the roller guides by determining which guide you need to realign. Push Play to load the tape. While the unit is running, try to turn each roller guide post. If one post can be turned easily while the other can't, you need to realign the one that you could turn.

Checking the adjustment

Connect the VCR to the TV monitor and check the adjustment of the roller guides. Insert a high-quality rented video tape, which has been produced by a well-known production company. Now, push Play. If the roller guides are properly adjusted to the tape edge guideline, you'll have a perfect picture. If not, you might experience one of two things. A thin belt of lines might run horizontally across the top or bottom of the picture. On the other hand, the picture might jump up and down or have a jitter. If either of these problems occur, fine-tune the roller guides.

Fine-tuning the roller guides

When fine-tuning the roller guides, use only a high-quality rented video tape. If you have any lines across the bottom of the picture (Fig. 20-14), insert the proper tool into the take-up roller guide and turn it clockwise until all the lines disappear off the bottom of the screen. When you reach this point, stop and make a mental note of the position. Then, continue in the same direction for another ¼ turn and stop. Remove the tool. The take-up roller guide is now fine-tuned.

If you have any lines across the top of the picture (Fig. 20-15), insert the proper tool into the supply roller guide and turn it clockwise until all the lines disappear off the top of the screen. When you reach this point, stop and make a mental note of the position. Then, continue in the same direction for another ¼ turn and stop. Remove the tool. The supply roller guide is now fine-tuned.

When fine-tuning the tape path alignment, if more lines appear when turning the roller guide clockwise, the problem is the positioning of the video tape edge guideline is incorrect. Go back to the section in this chapter on "Adjusting the roller guides."

You might find jittering in the picture after you've adjusted the video tape to the tape edge guideline or after you've fine tuned it. To correct this problem, insert the proper tool into the supply roller guide and turn it clockwise approximately ¼ turn. Then, fine-tune the adjustment back and forth, approximately ⅛ turn until the jittering stops. If this adjustment doesn't completely eliminate the jittering, you can fine-tune the take-up guide to completely stop the jittering. You shouldn't have to turn it more than ⅛ turn. This final adjustment will correct the problem.

If you turn on the TV monitor and find a confusing number of lines, then both roller guides are out of alignment. Return to the tape edge guideline adjustment and readjust from

20-14 Lines across the bottom of the picture.

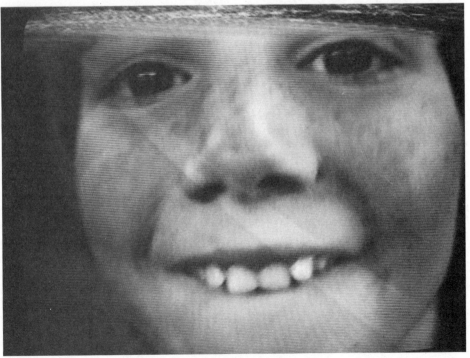

20-15 Lines across the top of the picture.

that point. If the tape edge guideline has been properly adjusted, turning the take-up guide clockwise will make the lines go down and off the screen. Turning the supply guide clockwise will make the lines go up and off the screen.

These adjustment procedures are correct for all two-hour video cassettes, if you are using a rented professionally-produced video tape.

Fine-tuning roller guides for six-hour tapes

When inserting a six-hour video cassette after you've fine-tuned for a two-hour cassette, you might find a thin line or lines in the upper or lower portion of the picture, as shown in Fig. 20-16. In this case, you'll need to fine-tune for the six-hour tape. Insert a video tape recorded by your machine when you first purchased your unit. The unit was aligned properly at that time; consequently, the video tape also will be aligned. Simply fine-tune the roller guides as previously explained.

After aligning the roller guides and inserting a video cassette recorded just prior to the alignment procedure, you'll probably find that the tape now has lines in the picture, but all other video cassettes function properly. All previously recorded tapes were recorded when the unit was out of alignment. When viewing these tapes, you'll have to adjust the tracking control to one end or the other or rerecord the tapes.

20-16   Fine lines in the picture.

Tightening the base lock nuts

The last step is to tighten the lock nuts on the roller guides. Locate the roller guide lock nuts, insert the proper tool, and turn it clockwise until the nut is snug. Do not over tighten the nut. Overtightening can cause the roller guide to turn, and you'll lose the alignment.

Review

- ☐ Center the tracking control.
- ☐ Loosen the base lock nut before turning the roller guides.
- ☐ Make sure the roller guides turn smoothly.
- ☐ Make sure you don't turn the base lock nut too far clockwise, or the roller guides will turn by themselves as the video tape passes around them.
- ☐ Don't touch the angle adjustment behind the V-mount.
- ☐ Adjust the bottom edge of the video tape to sit on the tape edge guideline.
- ☐ Fine-tune the roller guides (The take-up roller guide affects the bottom of the picture, and the supply roller guide affects the top of the picture.)
- ☐ Tighten up the base lock nuts.

21

The angle adjustment

THE ONLY UNITS WITH AN ANGLE ADJUSTMENT HAVE A LOCK NUT ON TOP OF the base of the roller guides. There's one angle adjustment located behind the V-mount, as shown in Fig. 21-1. These factory-set angle adjustments are critical and shouldn't be touched. This chapter is only for those of you who have already turned the adjustments before reading this book. If this adjustment has been turned over half a turn clockwise, a spring located under the V-mount gets bent. The roller guide will be tilted, and you'll not be able to perform the video tape path alignment.

Removing the angle adjustment assembly

If you've turned this adjustment, the leaf spring under the V-mount is bent. Take out the screw directly behind the angle adjustment, as shown in Fig. 21-2. Now, remove the V-mount and pick up the spring (Fig. 21-3). The place the spring gets bent is shown in Fig. 21-4. Bend the spring back, so when looking at its profile it is flat. You'll find a slight upward edge on the tip of the center piece of the spring. Leave it, its suppose to be this way.

Remounting the angle adjustment assembly

Holding the V-mount with a 1.5mm allen wrench, adjust the angle adjustment nut counter-clockwise until it is flush with the bottom of the V-mount. Put the spring back onto the transport. Line up the two pin holes on the spring with the holes on the transport, keeping the lip on the center piece pointing up. There are two pins on the bottom of the V-mount, as shown in Fig. 21-5. Place the V-mount on top of the spring, making sure these pins go through the holes in the spring and into the transport. Replace the mounting screw and tighten it down.

21-1   The angle adjustment.

21-2   The mounting screw to the V-mount.

Adjusting the angle adjustment

Take a 1.5mm allen wrench and place it on the angle adjustment nut. Slowly turn the angle adjustment clockwise. You'll notice the allen nut turns very easily. Stop when you feel it get snug against the metal spring. Now, insert and load a video tape. Starting with the side you just repaired, look at the tape guide post. See if the roller guide post is perfectly straight up

21-3 Removing the spring.

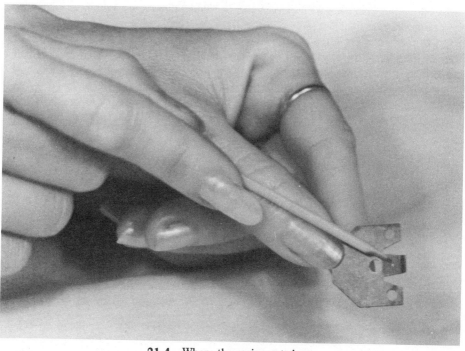

21-4 Where the spring gets bent.

21-5 The two pins on the bottom of the V-mount.

and down. At this point, the guide post should be slightly tilted toward the front of the unit. For a reference point, look at the top center for a circle in the other roller guide and check the position of the circle in the V of the V-mount (Fig. 21-6). Turn the angle adjustment to match the position of the other roller guide. You shouldn't have to turn the adjustment more than ½ turn. If both angle adjustments are out of alignment, adjust both roller guide posts straight up and down.

After completing the above procedure, the unit needs a video tape path alignment. Refer to chapter 20, "Video tape path alignment."

Review

☐ Do not touch the angle adjustment unless you have already accidentally turned it.

☐ If you have turned this adjustment, the leaf spring under the V-mount is probably bent.

☐ Remove the V-mount and spring.

☐ Straighten out the leaf spring.

☐ Turn the angle adjustment all the way up, before remounting.

☐ Remount the leaf spring and V-mount.

☐ Run the angle adjustment down until it is snug against the spring.

☐ Load a video tape while checking for a straight fine line in the video tape on both sides of that particular roller guide.

☐ Now, you can do a video tape path alignment.

21-6   The top position of the roller guide post in the V-mount.

22

Audio head alignment

THE AUDIO HEAD IS INCORPORATED INTO THE A/C HEAD. "A/C" STANDS FOR audio control. Usually a weak or distorted audio signal is caused by a dirty or improperly aligned A/C head. A buildup of residue prevents the video tape from making flush contact with the A/C head, producing a weak or distorted audio signal. Refer to the section in chapter 4 on "A/C head" under the subhead "Cleaning the various components." If you have a stereo hi-fi model, you'll have to put the stereo hi-fi switch in the normal position before proceeding with this alignment. If you have a stereo switch, place it into the stereo position.

Ideal setting for an A/C head

Figure 22–1 is a diagram of the front of an A/C head. On all A/C heads, you'll find three sets of fine parallel lines or small metallic rectangular squares. Beginning with the top left side, the first metallic square is the audio erase head. Straight across, on the upper right side, is the audio playback and record head. The third metallic square is at the bottom right side. It is the FM head. The distance between the center of the audio head and the center of the FM head is the exact width of a video tape. The ideal setting for an A/C head is straight up and down, not tilted. The top edge of the video tape should barely touch the center line of the audio and erase head, as shown in Fig. 22–2.

Location and function of adjustment screws

A/C heads are mounted on a set of brackets, as shown in Fig. 22–3. The upper bracket has three adjustment screws and fastens directly to the A/C head. The three adjustment screws go down through the upper bracket to the lower bracket. The upper bracket is mounted to the lower bracket by these three screws making the upper bracket adjustable.

Figure 22–4 shows the upper bracket and provides a better view of the three adjust-

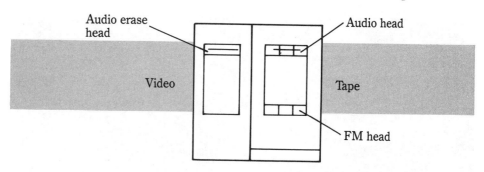

22-2 The ideal setting for an A/C head.

ment screw holes. Models manufactured from approximately 1979 to 1983 use this type of mounting bracket. When this bracket is mounted in its proper position, you'll find an adjustable set screw going through each screw hole.

The first set screw is on the right side of the A/C head and has a spring wrapped around the screw above the bracket. This spring-loaded screw keeps tension on the A/C head. You shouldn't have to adjust this screw unless you're replacing the A/C head. In this case, all you need to do is remount the screw with the spring on it. Adjust the screw down until there is tension on the A/C head. When adjusted properly, the A/C head is stiff but remains flexible.

The second screw is located at the rear of the A/C head and usually will have a 1.5mm allen nut, but a few models have a Phillips head screw. This adjustment screw tilts the A/C head forward or backward. This screw is used for the tilt adjustment.

22-3   The A/C head mounting brackets.

22-4   The three adjustment screw holes.

22-5 The height adjustment.

The third set screw is located on the left side of the A/C head and is a Phillips head screw. This screw is for making the azimuth adjustment, which allows you to change the angle of the A/C head from side to side.

The last adjustment screw is directly behind the A/C head. This nut controls the height adjustment. You'll find a metal shaft with a nut on top of it directly connected to the bottom bracket, as shown in Fig. 22–5. This adjustment nut moves the A/C head and both brackets up or down. To adjust this, nut you'll need a 7/32-inch socket, a 1/4-inch socket, or a nut driver.

In some models the A/C heads adjustment screws are in the same location, except the spring loaded screw is upside down and comes up from the bottom of the mount.

In most models, you'll notice a little screw and shield plate mounted on top of the A/C head, as shown in Fig. 22–6. Remove the shield plate, so you have a clear view of the front of the A/C head.

Tilt adjustment

Refer to Fig. 3–1 to reorientate yourself with the terminology of the video tape path. Insert a video cassette and push Play. Look straight down on the video tape to see if it looks like a perfectly straight hairline between the take-up roller guide and the A/C head. Figure 22–7 shows you where to look. Also check to see if the video tape is being wrinkled at the bottom protruding edge of the take-up tape guide. When the A/C head is adjusted properly, the video tape on the take-up tape guide should look like Fig. 22–8.

If the video tape has a twist coming from the roller guide or is being wrinkled at the take-up tape guide, place a 1.5mm allen wrench into the adjustment screw and slowly turn it counterclockwise until the twist or wrinkle disappears. Then, turn the adjustment clockwise to move the top of the A/C head forward toward the video tape. At the same time, look

22-6   The shield plate and its screw.

22-7   The point to look straight down at the video tape.

straight down on the video tape where it comes off the take-up roller guide and makes contact with the A/C head. Make sure the video tape doesn't get twisted. If you go too far with this adjustment, it causes the video tape to wrinkle on the bottom protruding edge of the tape guide. If this happens, slowly turn the adjustment counterclockwise until the wrinkle disap-

22-8 The protruding edge on the take-up tape guide.

pears. You want to move the A/C head as far forward as possible without leaving a twist in the video tape. If the twist still remains, tighten the torque adjustment one or two notches.

Height adjustment

Locate the height adjustment nut on top of the metal shaft directly behind the A/C head. Find the proper size socket or nut driver and place it in the nut. Turn the nut counterclockwise while keeping an eye on the front of the A/C head. Watch for the metallic square on the face of the A/C head to come up above the video tape, as shown in Fig. 22–9. If you have the TV monitor on at this time, you'll find the picture has a belt of horizontal lines running up through it. Do not worry, the lines will disappear after doing the audio alignment.

Azimuth adjustment

Insert a Phillips head screwdriver into the azimuth adjustment, located on the left side of the A/C head. This adjustment tilts the A/C head from side to side. Adjust the A/C head, so the top two metallic squares are horizontally parallel to the top edge of the video tape. These two metallic squares are the audio eraser head and the audio playback head.

Now, go back to the height adjustment nut. Turn the nut clockwise. Move the A/C head down until the top edge of the video tape just covers the center line in the middle of the two metallic squares. Figure 22–2 shows how the video tape should look up against the A/C head.

22-9   Adjusting the metallic squares above the video tape.

Fine-tuning the audio head

You'll need to fine-tune the audio head once the other adjustments are completed. Insert a rented video tape that was produced by a major production company. Approximately 30 seconds to 5 minutes after the movie has ended, the picture will be a white rastor, or a white blank screen. During this time, you'll hear a 1000 note cycle audio tone that lasts up to five minutes. This audio tone is a wake-up call for anyone who fell asleep during the show. You're going to fine-tune the A/C head using the audio tone. Locate the audio signal on the tape and turn up the audio on the TV monitor.

Fine-tune the height adjustment by turning the screw back and forth slightly, listening to the tone. Determine the point at which the volume peaks. Set the adjustment there. You shouldn't have to turn this adjustment more than ¼ turn in either direction.

Locate the azimuth adjustment on the left side of the A/C head. Use a Phillips head screwdriver to turn the adjustment screw in either direction and listen to the tone. You should hear the audio tone increase or decrease in volume. Be sure to turn the adjustment at a moderate speed. You want to turn the adjustment in the direction that causes the volume increases. At some point, the audio signal starts lowering in volume. Keep turning the screw back and forth until you find the loudest point and leave it set there.

Locating the azimuth adjustment in newer VCRs

This section is for models made approximately after 1983. You will follow the exact procedure for A/C head alignment, with two exceptions. First, the spring-loaded screw on the

22-10 Locating the azimuth adjustment in newer VCRs.

right side of the A/C head has been switched with the screw on the left side. In other words, the azimuth adjustment has been moved from the left side of the A/C head to the right side, as shown in Fig. 22–10. You'll usually find red paint on the side of the azimuth screw. The paint locks the screw down, keeping the screw from turning on its own. Follow the same procedures as previously explained to align the azimuth adjustment. After aligning the A/C head, place red nail polish on the side of this screw to lock it down. The second change is that the tilt adjustment, or the screw located directly behind the A/C head, has been changed from a 1.5mm allen nut to a Phillips head screw.

Audio alignment on early VCRs

Early VCRs used a different method to align the A/C head. These models have three set screws, but no height adjustment nut. All the set screws are located in the same place; the difference is that they're all spring loaded, as shown in Fig. 22–11. (I've picked the A/C head, and brackets up to give you a better view.) Adjust the height of the A/C head by turning all three screws, one after the other. Turn them counterclockwise to move the head up or clockwise to move the head down. Count each turn you make on the first screw. Turn the other two screws the same number of turns as the first to keep the audio head straight up and down.

Turning the screws counterclockwise causes the metallic squares to rise on the face of the A/C head above the top edge of the video tape. As you bring the head back down, you'll need to adjust each screw accordingly. Keep the top edge of both metallic squares parallel to the top edge of the video tape. Adjust the metallic squares down to the center lines, as previously explained.

The same adjustment screws are used for another purpose. The adjustment screw located at the back of the A/C head is still the tilt adjustment. The adjustment screw on the left side of the A/C head is still the azimuth adjustment. Follow the proceeding procedure for the audio alignment. Although this model method is harder to align, it maintains the alignment longer.

Aligning newer VCRs using the old method

Newer VCRs have the same spring-loaded set screws as the older units. (See Fig. 22–11) The tilt adjustment, however, has been moved from behind the A/C head to in front of it. The mounting is still the same. Adjust and align the A/C head as previously explained in the older units.

22-11   An A/C head mounted on spring-loaded screws.

A tip on recording bad audio

Upon playback of a personally recorded video cassette, the picture is okay, but the audio portion is distorted or weak. This condition could be caused by a dirty A/C head. The head was just dirty enough to keep the audio signal from imprinting properly on the video tape. On other prerecorded tapes, the audio signal is okay. Clean the A/C head and record the tape again.

A tip on adjusting the FM portion

If you have lines floating from the bottom to the top of the picture, the quickest way to correct this problem is to adjust the height adjustment. Adjust the height adjustment down, then move it up until the top edge of the video tape just covers the center line in the top metallic boxes on the front of the A/C head. The problem is that the video tape is either too high or too low on the A/C head, so the FM portion of the A/C head is unable to read the proper signal off the video tape. This method is a short cut. The proper procedure is to go through the entire alignment process.

Review

☐ Clean the A/C head.

☐ There are four audio head adjustments.
- The screw on the right is for putting spring tension on the A/C head.
- The screw in the back, called the tilt adjustment, is for taking wrinkles and twists out of the video tape.
- The screw on the left is the azimuth adjustment, for peaking out the audio signal.
- The nut on top of the shaft behind the A/C head is the height adjustment and is used to line the top edge of the video tape to the center line in each metallic square located near the top of the A/C head.

☐ Adjust the tilt adjustment to allow the most pressure against the video tape without twisting or wrinkling the tape.

☐ Adjust the height adjustment, so you can see the metallic square on the face of the A/C head above the video tape.

☐ Adjust the azimuth adjustment to make the metallic squares on the A/C head parallel to the top of the video tape.

☐ Adjust the height adjustment down to where the top edge of the video tape is just covering the center of both metallic squares located near the top of the A/C head.

☐ Adjust the azimuth adjustment using an audio tone for peaking out the audio signal.

23

Adjusting the tracking control

A VIDEO TAPE CARRIES THREE SEPARATE SIGNALS. FIGURE 23–1 SHOWS WHERE the signals are put on a video tape. The top portion of the video tape carries the audio signal, which is adjusted by doing an audio head alignment. The middle portion carries the video signal, which is adjusted by doing a video tape path alignment. The video signal is produced on tracks. The bottom portion of the video tape carries the FM signal.

The FM alignment centers the tracking control. If the audio and video portions are functioning properly (no stationary lines in the picture), you can proceed with this adjustment.

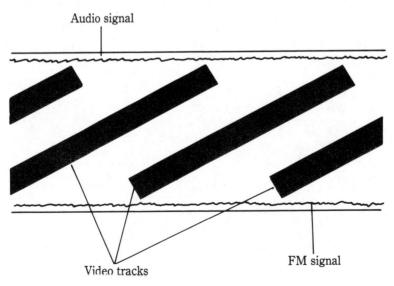

23-1   The three signals on a video tape.

Checking the tracking control

Insert a rented video tape and push Play. Find the tracking control on the front of the unit. Place the tracking control to its center position. There should be an indentation or a niche. Rotate or slide the tracking control knob to the right until lines appear either from the top or the bottom in the picture or until the picture becomes completely distorted with lines. Return the tracking control to the center position and note the distance you traveled to the right before the lines appeared. Now turn or slide the tracking control knob to the left until the lines or distortion appear in the picture, again noting the distance back to the center position. You need to do this to see if all the lines start at the same point in either direction from the center.

If the adjustment isn't centered properly, you'll be able to adjust the tracking control all the way in one direction without showing any lines from the center point, but in the opposite direction, lines appear right away. On the other hand, you might find the tracking control is set at one end or the other in order to get a clear picture. To correct this problem do an FM alignment. If you have a unit with push-button tracking controls, don't make this adjustment as long as lines appear in both directions.

Locating the FM adjustment

In most models, the FM alignment adjustment nut is located at the front right corner of the A/C head mounting bracket, as shown in Fig. 23-2. In a few models, this adjustment is located to the left side of the A/C head. The FM alignment adjustment nut is the same no matter what model you have. It'll be either a large brass or silver adjustment nut with slots on top and on each side of the threaded shaft its mounted to. In newer models, the FM adjustment is located behind and to the left of the A/C head, as shown in Fig. 23-3. To adjust this nut place a medium flathead screwdriver in either slot. (See Fig. 23-4)

A few models don't have an adjustment nut. In these models, the A/C head is mounted with spring-loaded screws between the upper and lower A/C head brackets. The lower bracket is mounted directly to the transport by two screws, one on each side of the A/C head. Loosen the mounting screws. The entire assembly now can slide from front to back for adjustment.

Centering the tracking control

Connect the TV monitor and turn it on. Insert a rented or purchased video tape, then push Play. The tracking control is set at its center position or niche, and the picture is okay. As you adjust the tracking control in one direction, you should find no noticeable change in the picture. Now, return to the center position. Adjust the tracking control in the opposite direction. Lines should appear soon after you start to turn the tracking control, indicating that the tracking control isn't centered. Return the tracking control to its center position. Slowly adjust the tracking control knob back in the direction showing the lines. Continue until three quarters of the picture is covered with lines (Fig. 23-5). Leave the control set it at that point.

Place a flathead screwdriver into the FM alignment adjustment nut and turn it in either direction. As you turn it in one direction, you should notice the picture getting worse. When

23-2 First location of the FM adjustment nut.

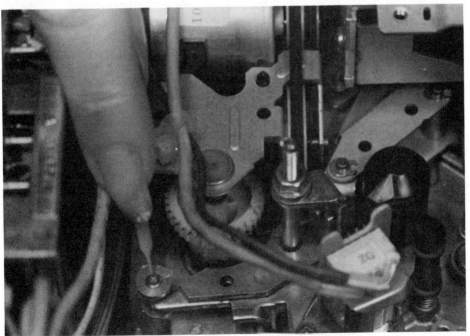

23-3 Second location of the FM adjustment nut.

23-4 Adjusting the FM alignment nut.

you turn it in the opposite direction, the picture will start to clear up. Continue in this direction until the entire picture is clear and then stop. In most models, it'll take about three quarters of a turn, but in some models, it can take up to two complete turns to remove the lines. Now, place the tracking control back to its center position. Turn or slide the tracking control knob in both directions, to see if the lines start to appear at the exact same point from the center. If it isn't centered, then adjust the tracking control in the direction showing the lines when the control is the closest to the center position or niche.

In the models with no adjustment nut, slide the entire assembly in one direction or the other. Sliding the assembly is how you make the FM adjustment in these models. Adjust the tracking control until the lines appear, then slide the assembly slowly to remove the lines. Follow the same procedure explained in the previous section for centering the tracking control.

To fine-tune the FM adjustment, insert a blank video cassette and make a recording using the six-hour mode. Play it back and check to see if the tracking control is centered properly. If it isn't follow the same procedure for aligning this tape. You shouldn't have to turn the adjustment more than ⅛ or ¼ turn.

23-5 Lines in the picture.

Tracking adjustment is completely to one end

If the tracking control must be adjusted all the way to one end or the other to receive a clear picture, follow the following procedure to get the tracking control back to its indentation or niche. From the point the tracking control is set for a good picture, adjust the tracking control back until you receive lines covering three quarters of the picture. Now, turn the FM adjustment nut to clear the lines, as previously explained.

Repeat this procedure until the picture is clear when the tracking control is in its centered position or niche. You shouldn't need to turn the adjustment more than two or three complete turns. Now, follow the instructions in the previous section to finish centering the tracking control.

If, after making the proper adjustment, you find the tracking control is off again, then the tape was recorded prior to doing the FM alignment. The FM alignment isn't at fault. The problem is that the video tape was recorded on a misaligned unit. If this condition occurs, do not realign the FM adjustment nut, instead refer to chapter 20, "Video tape path alignment."

Adjusting VCRs without an FM adjustment nut

To determine if you have this type of model, look at the A/C head to see if it's mounted with spring-loaded screws. These springs are located between the upper and lower mounting bracket of the A/C head, as shown in Fig. 22-11. The bracket might look different, but the springs are the same. To make this adjustment, loosen the two screws holding the bottom bracket to the transport. Slide the complete A/C head assembly either toward the front or the back of the unit. Use the same procedure as previously explained. You'll be using the A/C head itself to make the lines disappear, instead of the FM adjustment nut.

Review

☐ Only use a rented or purchased video tape for making tracking adjustments.

☐ Use the tracking control knob to see if you have lines appearing in the picture at equal distances from the center position.

☐ If lines appear in only one direction, you need to adjust the FM adjustment nut.

☐ Turn the tracking control until the lines appear in the picture and adjust the FM adjustment until they disappear.

☐ Recheck the tracking control in each direction, for equal distance from the center position when the lines appear in the picture.

24

Replacing
the audio head

CONNECT THE VCR TO THE TV MONITOR AND TUNE IN A LOCAL BROADCASTED channel on the VCR. Check the audio portion of the program. If the audio portion is functioning properly, then insert a video cassette and push Play. If the picture is clear but the audio is malfunctioning, you have to determine if the A/C head is out of alignment, the audio head is bad, or the audio circuit is not working. Remember the audio head is incorporated in the A/C head.

Checking for a bad audio head

Push Play and turn up the volume control on the TV monitor. If you have a stereo hi-fi model, put the stereo switch into the normal position. Using the flat surface of a small flathead screwdriver, gently tap on the face of the A/C head while the video tape is playing, as shown in Fig. 24-1. Tap on the top right corner of the A/C head. You should hear a popping sound coming from the speaker in the TV monitor. For reference on the audio portion of the A/C head, refer to the section in chapter 22 on "Ideal setting for an A/C head."

If you hear a popping sound, the video tape isn't making a flush contact with the A/C head. This condition occurs for two reasons. The first reason is a buildup of residue on the A/C head. In this case, just clean the A/C head. Refer to the section in chapter 4 on "A/C head" under the subhead "Cleaning the various components." The second reason is that the face of the A/C head might have a small dent in it. These dents can occur when the A/C head is improperly cleaned. The surface of the A/C head dents easily. Don't use a hard or rough object to clean off the residue buildup. The dent causes the video tape to make improper contact with the audio portion of the A/C head. If an A/C head has been dented, it'll have to be replaced.

If you hear the popping sound, go to the small circuit board connected to the back of the A/C head. Touch the tip of a screwdriver to each contact directly behind the audio portion of the A/C head. (A contact is a soldered connection. A connection is where two or more com-

210

24-1 Tapping on the top right side of the A/C head.

ponents are soldered together.) The two top left pins protruding through the circuit board behind the A/C head are the contacts to touch. (See Fig. 24-4) Touch each contact one at a time, as shown in Fig. 24-2. *Do not short any of the contacts together.*

What you should hear when coming across the correct contact is a loud buzz coming from the speaker. You must have your finger on the screwdriver to create the buzzing sound. If you do hear a buzz, the audio circuit is working. You'll find that one or two contacts will cause this buzz. The buzz should be louder than the normal volume. If you do have a buzzing sound but no popping from the face of the A/C head, the A/C head has gone bad. If there's no buzzing sound, the audio circuit has malfunctioned. You'll have to take the unit into a local service center for circuit repairs.

To determine if the A/C head is completely out of alignment, use the tapping method on the face of the A/C head. Listen for the popping sound. This sound tells you the A/C head is functioning. Next, look at the metallic squares on the face of the A/C head. If you can see these squares while the unit is playing, the A/C head is completely out of alignment. Refer to the section in chapter 22 on "Ideal setting for the A/C head."

Replacing an audio head

Unplug the unit. Remove the two Phillips head screws located on the A/C head mounting bracket, one on each side of the A/C head, as shown in Fig. 24-3. One of these two screws has a spring wrapped around it. Remember which one has the spring and how to reassemble

24-2   Touching each contact on the back of the A/C head.

24-3   Mounting screws on the A/C head bracket.

24-4   Pins protruding out the back of an A/C head.

it. Do not remove the tilt adjustment screw or allen nut on the same mounting bracket directly behind the A/C head. This adjustment is used for realignment. After removing the two screws, grab the A/C head and lift it straight up. The mounting bracket will come up with it. Turn it around and look at the circuit board mounted on the back of the A/C head.

I've taken an A/C head out of another unit, so you can see what the back of an A/C head looks like without a circuit board (Fig. 24-4). Notice there are six pins protruding out the back of the A/C head. These pins go through a small circuit board and are soldered to it. To remove the circuit board from the A/C head, remove the solder from each pin. In stereo models, the A/C head has ten pins, instead of six, protruding out the back.

Using a solder wick to remove old solder

You'll need a soldering iron or soldering gun and some solder wick. Look at the circuit board where the pins come through. Place the solder wick over one of the soldered pins. Then, take the hot soldering iron and place it on top of the solder wick, as shown in Fig. 24-5. As the wick gets hot, it will absorb the old solder. Remove the iron and move the solder wick forward, to a clean spot on the wick. Place the hot iron back onto the wick to soak up the remaining old solder. You might have to repeat this procedure two or three times to remove all the solder. Do each pin the same way. In some models, you might find a wire attached to the same connection as the pin. In this case, touch the hot soldering iron to the connection and pull the wire off, as shown in Fig. 24-6. Then, use the solder wick to remove the old solder. It's helpful to draw a diagram. The diagram should show the color of the wire and which pin it was connected to. There will be no more than one wire and one pin soldered to

24-5 Using solder wick to remove old solder.

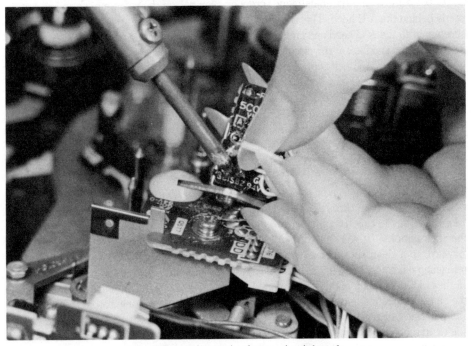

24-6 Removing a wire from a circuit board.

24-7   Removing the A/C head from the circuit board.

any one connection. You usually shouldn't have to remove more than two wires from the back of the circuit board. All the other wires don't have to be removed; they're connected to their own connections on the circuit board.

After removing the solder from each pin, simply pull the A/C head off the circuit board, as shown in Fig. 24-7. Take the A/C head, with the make and model number of the VCR, to a local electronic supply center for replacement, then proceed to the next section.

Soldering pins to a circuit board

Take the new A/C head and push the protruding pins through the pin holes on the circuit board. Take a hot soldering iron or gun and coat the tip with solder (Fig. 24-8). After coating the tip, tap the reserve solder off the iron onto a block of wood. Don't tap the reserve solder over the VCR. Position the tip of the iron to touch the side of the pin and the circuit board simultaneously. Push the solder into this connection until it flows all the way around the pin, as shown in Fig. 24-9. Remove the solder and leave the iron on the connection for another two or three seconds to bond the solder to the circuit board and the pin. Remove the iron and repeat this procedure for each pin.

To replace the wire to the pin connection, wrap the exposed end of the wire around the pin first. Then solder it, as previously shown.

Don't solder two pin connections together. They must be separate. If you put too much solder on a pin, the solder might flow over and connect to a neighboring connection, creating a shorted circuit. Simply use the solder wick to absorb any excess solder.

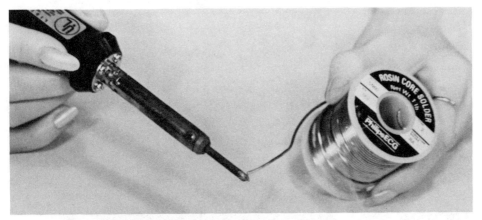

24-8 Coating the tip of the iron with solder.

24-9 Soldering a connection.

In some newer models, it's easier to remove the A/C head. The difference is a white plug located on the top near the back of the A/C head. In this case, unplug the plug, as shown in Fig. 24-10, and remove the A/C head, as previously shown for replacement.

Remounting the audio head

Locate the tilt adjustment screw in the old A/C head mount. The screw is located behind the A/C head. Measure the distance the set screw protrudes through the bottom of the mount,

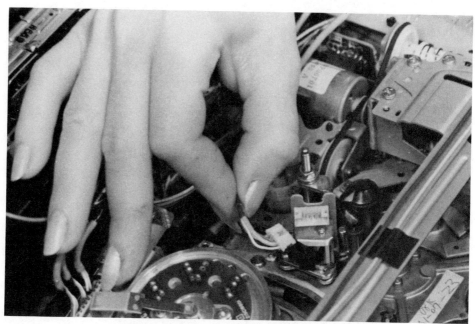

24-10 Removing the plug from the A/C head.

as shown in Fig. 24-11. This screw varies in length, shape, and type. Remove the screw from the old mount and place it in the new mount. Adjust the screw to the same measured distance as the old mount. This step will give you a good starting point for realigning the A/C head.

Place the A/C head in its original position. Replace the spring that was wrapped around the set screw before replacing it (Fig. 24-12). Replace the two set screws into the mount, one on each side of the A/C head. Adjust these two set screws, so the A/C head sets as level as possible. Now, you're ready to do an audio head alignment. Refer to chapter 22, "Audio head alignment."

Replacing an audio head with spring-loaded mounts

Spring-loaded mounts use three spring-loaded adjustment screws, instead of one, as shown in Fig. 22-11. These springs are located between the top and bottom bracket under the A/C head. For a reference point, measure the distance from the transport to the bottom of the A/C head, as shown in Fig. 24-13. Remove all three screws and springs to remove the A/C head. When you install the new A/C head, put the springs back onto each adjustment screw between the brackets. Measure the distance from the transport to the bottom of the A/C head. If the distance has altered, readjust all three screws. The measurement should be the same as before. The A/C head should be sitting level. Now, you're ready to do the audio head alignment. Refer to the sections in chapter 22 on "Audio alignment on early VCRs" and "Aligning newer VCRs using the old method."

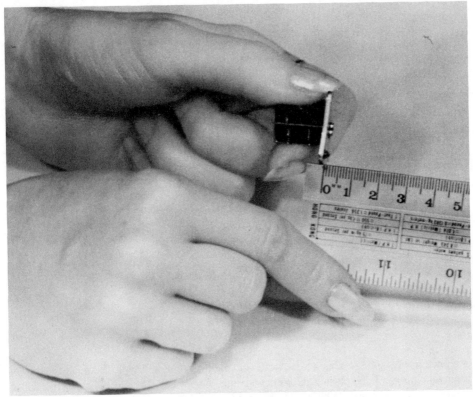

24-11   Measuring the tilt screw adjustment.

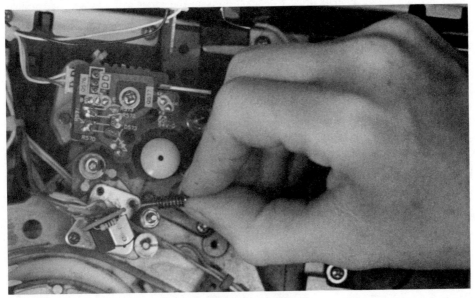

24-12   A spring-loaded screw.

24-13 Measuring the distance between the A/C head mounting bracket and the transport.

Review

☐ Place the unit into the play mode with a video tape inserted.

☐ Tap lightly on the audio portion of the A/C head to see if the speaker makes a popping sound.

☐ If it doesn't pop, touch the audio connections on the back of the A/C head. Listen for a buzzing sound coming from the speaker.

☐ If you hear a buzzing sound but no popping sound, the A/C head is bad and needs replaced.

☐ To remove the A/C head, remove the two screws on each side of the A/C head, leaving the tilt adjustment screw.

☐ If you have three spring adjustment screws, measure the distance from the transport to the bottom of the old A/C head before removing the bad A/C head.

☐ Remove the old solder from the circuit board with the aid of solder wick.

☐ Place the pins of the new A/C head through the circuit board and solder them in.

☐ Be sure you haven't soldered two connections together.

☐ Measure the distance of the tilt adjustment screw on the old A/C head mount and place that set screw into the new mount at the same distance.

☐ Remount both screws on each side of the A/C head and adjust these screws to make the A/C head as level as possible.

☐ Refer to chapter 22, "Audio head alignment."

25

Getting into the undercarriage

THE UNDERCARRIAGE CONTAINS A SYSTEM OF GEARS, MOTORS, BELTS, AND pulleys. To service and clean the undercarriage, remove the bottom cover plate, then open up the large circuit board. This gives you clear access to all the systems in the undercarriage.

Removing the bottom cover plate

Unplug the VCR. To remove the bottom cover plate, place the unit on its side. Remove all the Phillips head screws on the bottom cover plate. You'll find from 3 to 12 mounting screws on the bottom cover plate. In some models, once you remove the mounting screws, the plate will lift right off. For most models, pull the plate straight forward about ¼ inch to clear the plastic tabs holding the plate on. In a few models, pull the plate straight back about ¼ inch. Lift off the bottom cover plate. Some models have four rubber feet, one in each corner on the bottom cover plate. In the center of each rubber foot is a large black plastic Phillips screw. Remove the screws and each foot. This cover plate comes off. When remounting the bottom cover plate, reverse the process.

Bottom circuit board

After removing the bottom cover plate, you'll find a large circuit board covering the under-carriage. You must open the circuit board to reach the undercarriage. All circuit boards are hinged and swing open like a door. Determine which end of the circuit board lifts out and which end is hinged. Most circuit board hinges are located near the rear of the unit. In a few models, they're located on the side of the circuit board. Most of the circuit board mounting screws are located near the front of the unit. Some models have no circuit board covering the undercarriage. In these models, just remove the bottom cover plate.

Locating mounting screws

Each circuit board has 5 to 20 Phillips head screws. They aren't all mounting screws. The other screws serve different functions. You'll remove only the mounting screws to open the circuit board.

In most models, the mounting screws are at the opposite end of the circuit board from where you find the hinges. Mounting screws have special markings around the screw. The special marking is a white circle or a pointing arrow, a plastic washer under the screw, or the screws are different in color from the rest. For example, if you find metallic screws and red screws, the mounting screws are the red screws. If you find all red screws, then remove them all. The most common markings used for the mounting screws are a pointing arrow or red coloring.

Ways to open a circuit board

There are six different methods for opening the circuit board in your VCR, after you have removed the bottom cover plate. Read each of the following sections and follow the instructions listed in the section that pertains to your unit.

First way

The first way to open a circuit board is found in older models. On each side of the rear of the unit, you'll find two hinges attached to the circuit board. In this model, two long L-brackets act as hinges, as shown in Fig. 25-1. Usually with this type of hinge, there also will be a heavy

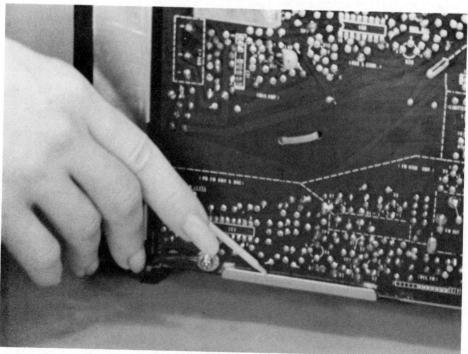

25-1   L-brackets used for hinges.

braided grounding wire. It'll be located beside each hinge, mounted by a screw to the circuit board. These grounding wires aid the hinges after opening the circuit board. Leave these wires and screws intact.

After locating the hinges, remove the circuit board mounting screws. There are five screws on this circuit board. Start by looking at the three screws located at the front of the unit, as shown in Fig. 25-2. Notice that the two outside screws have washers underneath, while the center screw doesn't. Remove the two outside screws. At the same end, pull open the circuit board like a door (Fig. 25-3). Now you can see the undercarriage with all its gears, belts, and pulleys. If the circuit board doesn't open easily, there's another screw to be removed. Don't use force on the circuit board, you can break it.

Second way

Look for the hinges at the rear of the unit. In this model, the hinges are two round white plastic cylinders. The cylinders are inserted into a slidable groove, as shown in Fig. 25-4. Next, locate the mounting screws. This model has three screws, with an arrow pointing at each one, as shown in Fig. 25-5. Remove only these three screws. At the bottom of the front cover, open the small panel doors (Fig. 25-6). I've removed the panel door to show the controls protruding through the front cover above the circuit board.)

If the controls are within ½ inch of the circuit board, the controls are attached. If any controls are near the circuit board, pull the board straight back approximately ½ inch, so the controls can clear the front cover, as shown in Fig. 25-7. Now, open the circuit board like a door (Fig. 25-8).

If the circuit board doesn't slide back and open easily, you have a different type of model than is shown. There might be another mounting screw to be removed.

25-2 Two mounting screws with washers underneath.

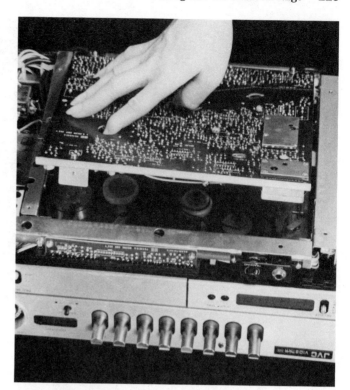

25-3 Opening the circuit board.

25-4 Locating the plastic cylinder hinges.

25-5   An arrow pointing a each mounting screw.

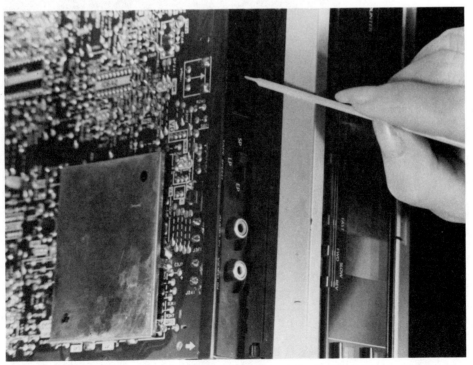

25-6   The controls protruding through the front cover.

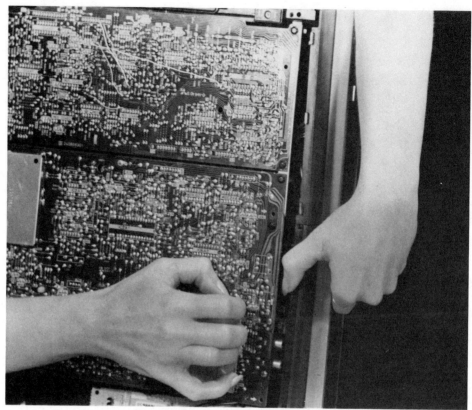

25-7 The controls clearing the front cover.

Third way

This model has two shafts made of a black plastic coming in from the chassis, one on each side of the circuit board. The hinge is a long metal bracket along the back of the circuit board. Each end of the bracket is inserted around the black plastic shaft, as shown in Fig. 25-9. Each mounting screw has a white circle painted around it (Fig. 25-10). These are the screws you need to remove. Check the front cover for a small panel door, as described in the previous section. If you have no panel door, lift up on the front of the circuit board.

Fourth way

This model has two to four white plastic hinges across the back of the circuit board, as shown in Fig. 25-11. The mounting screws are red in color and are near the front of the unit. Remove only the red screws. After removing the screws, you'll see four to six black plastic clips around the outside of the circuit board. The clips are the same as shown in Fig. 25-13, except that the base is the same width as the clip itself. You might find two black plastic clips on the outside and one or two in the middle of the circuit board, protruding through a small rectangular hole. To release the clips, pull back on the clip and simultaneously pull up on that portion of the circuit board. Release each clip until the circuit board opens.

25-8 Opening the circuit board like a door.

Fifth way

This model has two large circuit boards beside each other, which are held together with brackets. You'll see approximately 19 screws on these circuit boards. The screws are usually gold or black in color. One screw is red or blue and is mounted through the middle of the front bracket. Remove this different colored screw, as shown in Fig. 25-12. Next, look for any small panel doors on the front cover. Refer to Fig. 25-6 to visualize the controls protruding out the front cover. Pull the circuit board back to clear the controls and open the circuit board.

In a few models, there's a small black plastic clip sticking out through a metal bracket along the front of the circuit boards. The clip is in front of the screw you just removed. Push the clip down to release the circuit board. Then, pull the board straight back to clear the controls. The circuit board will open like a door.

Sixth way

If there's no small panel door on the bottom of the front cover, you've removed the mounting screws, and it still won't open, the front circuit board is attached to the bottom circuit board. First, remove the front cover. Proceed to the first five sections of chapter 35. After removing the front cover, you'll find four to six black plastic clips across the top edge of the front circuit board, as shown in Fig. 25-13. Push up on each clip one at a time, simultaneously pulling forward on the circuit board near that clip. After releasing all the clips, pull the bottom

25-9   A black plastic shaft and a metal bracket hinge.

25-10   A mounting screw with a circle around it.

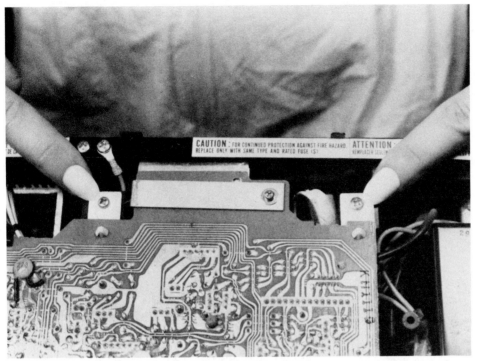

25-11   The white plastic hinges.

25-12   The red or blue mounting screw.

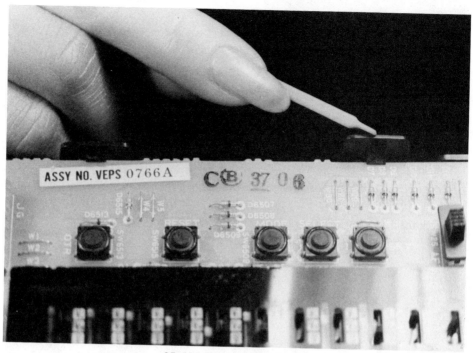

25-13 The black plastic clip.

of the front circuit board toward you about ½ inch. Moving the board will clear any tabs holding the front circuit board to the chassis. In this model, the front and bottom circuit boards open together like a door, as shown in Fig. 25-14.

If the circuit board still won't open

If, after removing the mounting screws, the circuit board still won't open, look on the circuit board for black plastic clips holding it down. If you start to open the circuit board and it bends in the middle, check each hinge for a locking red screw in front of the hinge. Figure 25-9 shows the type of bracket you would find these red screws in. Just remove the locking screws.

There are two more places to check if the circuit board still won't open. Either you have missed a mounting screw or someone has put the wrong colored screw in the wrong screw hole. If the screws have been switched, remove all the screws. Try to open the circuit board every time you take out a screw. It's important to remember which screw came from what hole. Each screw has a different thread, and you don't want to strip the threads.

Review

☐ Remove the bottom cover plate screw, then slide the plate forward for removal.

☐ To open a circuit board:
  • Look for the hinges located at the rear of the unit.
  • Look for the mounting screws, which are usually located at the opposite end from the hinges.

- The mounting screws are specially marked:
  - ~ a white circle around them
  - ~ an arrow pointing at them
  - ~ a plastic washer under them
  - ~ a different color from the rest

☐ Look for a small panel door located at the bottom of the front cover. If there is a door, you will need to pull the circuit board back before you can pull it open.

☐ After removing all the mounting screws, be sure to release any black plastic clips.

☐ Always check the bracket across the front of the circuit board for any clips that are keeping the board from opening.

☐ If the circuit board has no panel door and doesn't open after removing all the mounting screws, the front circuit board comes open with the bottom circuit board.

☐ Check the hinges for locking red screws located directly in front of the hinge.

25-14   The front and bottom circuit boards open together.

26

Servicing and lubricating the undercarriage

THIS CHAPTER GIVES YOU A COMPLETE LISTING OF ALL THE MOVING PARTS IN the undercarriage. Servicing and lubricating all the moving parts are very important for proper maintenance in a VCR.

Belt-driven VCRs

Belt-driven systems are still being used in some VCRs today. Get into the undercarriage and locate the main drive pulley mounted to the shaft on the capstan motor. The capstan motor drives the capstan shaft, fast forward, and rewind. You'll find one to four belts coming off this pulley. This pulley drives most of the moving parts in this model, as shown in Fig. 26-1. Note the belt configurations and remember their proper positions before you remove them. Each model might have a different configuration, but the basic idea is exactly the same.

I must stress the importance of placing belts back into their proper position. Notice the reel drive belt coming off the capstan motor pulley. This belt drives two separate pulleys and is in the correct position (Fig. 26-2). Now I'll give you an example of an incorrect position. After removing this belt, it would be easy to assume it would fit around the outside of the center pulley when replacing it, as shown in Fig. 26-3. If the belt is improperly placed onto the pulley, the center pulley will rotate backwards and the VCR will eat the video tape.

Look for any broken belts. If the fast forward or rewind systems aren't functional or the capstan shaft doesn't turn, these problems can be caused by a broken belt. Look in the undercarriage for any pulleys without belts. When a belt breaks, it falls off the pulley and down into the undercarriage. After finding the belt, it's time to go shopping. It's important to take the belt with you to the local electronics supply center. It also is helpful to know the make and model number of the VCR.

Remove all the belts from their pulleys. Start with the capstan motor pulley. First, remove the belt closest to you. Remove each belt separately as you work down the pulley. Except for the capstan drive belts and loading drive belts, all belts are easily removed. If you

26-1 The main drive pulley.

26-2 The correct position for the reel drive belt.

26-3   The incorrect position for the reel drive belt.

have difficulty in removing a belt, proceed to the section in chapter 27 on "Removing loading motors" or to the section in chapter 31 on "Checking a belt-driven VCR" and look at Fig. 31-5. After removing all the belts, clean the pulleys.

Cleaning pulleys

Saturate a glass brush in either cleaning alcohol or head cleaner. Place your finger on the opposite side of the pulley to hold it in place while cleaning. Place the glass brush in the groove of the pulley. Rub back and forth with a scrubbing motion, as shown in Fig. 26-4. Clean the brush, rotate the pulley, and clean the next section. Continue this procedure until you've cleaned all the way around the pulley. Clean each pulley in this manner.

Checking and cleaning belts

Take a clean paper towel and fold it in half. Saturate one corner of the towel in a cleaning solution. Place a belt on this corner of the saturated towel and fold the towel over the belt tightly, as shown in Fig. 26-5. Pull the entire length of the belt through the paper towel in short intervals. You'll notice a lot of black residue coming off of the belt. Periodically reposition the belt in the towel. Continue this procedure until there's no black residue coming off the belt. Clean all the belts in the same manner.

While cleaning each belt stretch it and release one end of it. If it pops back like a rubber band, the belt is still good. If it slowly returns to its normal size, the belt needs to be replaced.

26-4 Cleaning a pulley with a glass brush.

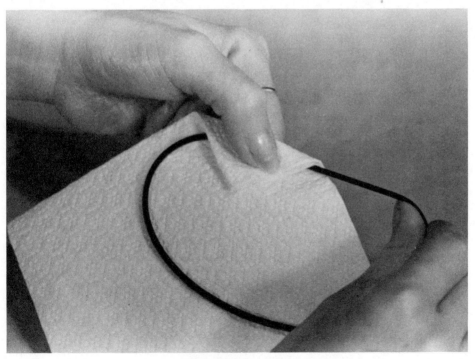

26-5 Cleaning a belt.

Checking and lubricating pulleys

Before replacing the belts onto their pulleys, check the pulleys for lubrication. Spin each pulley to feel if it moves freely. If it doesn't, the shaft of the pulley needs to be lubricated. If it is hard to turn the pulley, it has a buildup of residue between the shaft and the inside of the pulley arm. On the other hand, it could have lost its lubrication and became dry. Remedy this problem by spraying degreaser into the shaft at the point shown in Fig. 26-6. (I've removed a pulley assembly for demonstration purposes only.) After spraying the shaft, spin the pulley and continue this process until the pulley turns freely.

The next step is to lubricate the shaft of the pulley. Take a small flathead screwdriver and place a drop of household oil onto its tip. Place the dry side of the tip onto the shaft at the base of the pulley arm assembly, at the point where the shaft connects into the arm, as shown in Fig. 26-7. Now, tip the screwdriver up, allowing the oil to drop down onto the shaft. Go to the opposite end of the pulley arm assembly and lubricate this point in the same manner. Be careful not to get any oil on the small wheel attached to the shaft. Then, spin the pulley to work the oil up inside the arm onto its shaft. Repeat this procedure until the pulley spins freely.

Now, begin replacing the belts onto their proper pulleys. Make sure each belt fits securely and snugly. If the belt is loose or droops, then it is stretched out and has to be replaced.

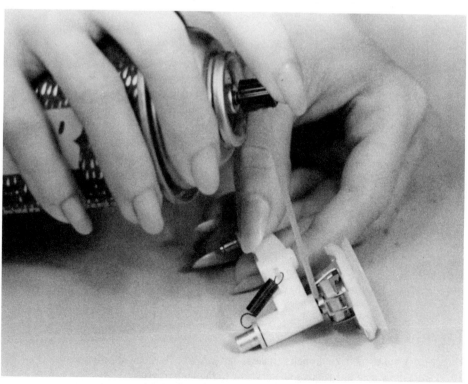

26-6 The point to spray degreaser.

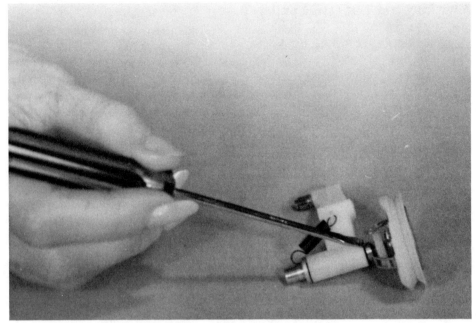

26-7　Placing a drop of oil onto the shaft.

Locating and lubricating all moving parts

Leave the VCR on its side, plug it in, and turn it on, so you can observe the moving parts inside the undercarriage. Insert a video cassette and push Play. Look for hinges connected with a C-, E-, or O-ring. Push Stop and observe all the moving parts and hinges. Push Play and Stop several times to study the points needing lubrication. Unplug the unit and place one or two drops of oil onto each hinge, as shown in Fig. 26-8. Don't get any oil on the belts or pulleys you just cleaned.

Gear-driven VCRs

Gear-driven systems use a combination of gears and have only one or two belts. Clean the belts and pulleys as previously explained. It's not necessary to clean the gears. To lubricate the gears, look for a C-, E-, or O-ring attached to the shaft in the center of the gear and place one drop of oil behind the ring onto the shaft, as shown in Fig. 26-9. Don't remove these gears. These gears are perfectly aligned to each other. Any alteration of the gears will cause the unit to jam. Don't try to spin or turn these gears as you did the pulleys.

Direct drive VCRs

Direct-drive systems have no belts, pulleys, or gears to be cleaned or lubricated (Fig. 26-10). Direct drive systems require no servicing to the undercarriage.

26-8   A lubricating point.

26-9   Lubricating the shaft of a gear.

26-10 A direct drive system.

Locating lubricating points that use grease

Leave the VCR on its side, insert a video tape, and push Play. As the unit loads, observe all the moving parts. You're looking for any bracket or lever that slides back and forth. Look for any moving bracket making a slight contact against the transport. Look for any rivets protruding through a slot in a bracket. The bracket moves back and forth leaving a trail of old grease, as shown in Fig. 26-11. Any place you find a trail of old black grease has to be relubricated.

Check the grooves on any cam gear located behind a small lever, as shown in Fig. 26-12. Cam gears also are located under a mounting bracket or on the side of a frontload cassette carriage. The cam gear has black grease in the grooves of the gear. It moves only when the unit is loading or unloading a video cassette or video tape.

The last place to check is the worm gears. Worm gears are directly connected to the loading motor. A loading motor drives the gearing to load or unload the video tape. A worm gear also can be located on the side of a cassette carriage, directly connected to the housing loading motor, as shown in Fig. 26-15. These are the only two places you'll find a worm gear.

After a period of time, the grease can fill up with dust and become gooey. The grease

26-11 A lubricating point that uses grease.

also can wear thin or become hard, causing the VCR to overwork. The effect is a slipping or broken drive belt.

Replacing old grease

To remove the old grease from a sliding bracket, just wipe it off with a paper towel. If your model has a worm or cam gear, place a small flathead screwdriver inside a paper towel and remove the old grease in the grooves, as shown in Fig. 26-13. A more favorable approach is to use a can of degreaser. Spray each point until it's free of old grease. Use a paper towel to absorb the residue.

The next step is to take some tuner grease or phono lube and place a small amount on the tip of a small flathead screwdriver. Spread the grease onto any moving bracket from which you removed the old grease. You also need to spread new grease onto the grooves of the cam gear (Fig. 26-14).

In models with a bracket over of the cam gear, you'll see only part of the gear sticking

26-12 The grooves in a cam gear.

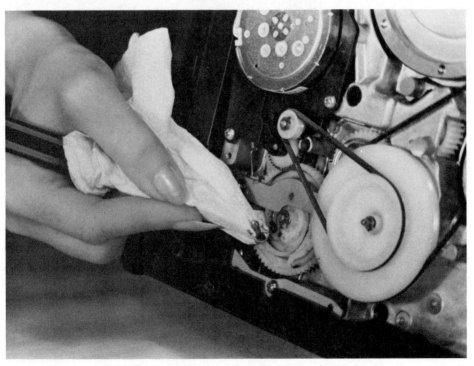

26-13 Removing old grease from the grooves in a cam gear.

26-14 Spreading new grease in the grooves of a cam gear.

26-15 Spreading new grease on a worm gear.

out. You'll need to push Play to make the gear move. While the gear is moving, place grease into the groove as it passes the open area. Push Stop to move the gear in the opposite direction. You might need to repeat this process several times to distribute the grease evenly.

To lubricate the worm gears, spread the grease on the gear, as shown in Fig. 26-15.

Insert a video cassette and push Eject to distribute the grease on the worm gear. If the worm gear is in the undercarriage, Push play and Stop to distribute the grease.

Review

- ☐ Check for any broken belts.
- ☐ Examine and remember the position of each belt before removing it.
- ☐ Be sure to clean all belts and pulleys.
- ☐ While cleaning the belts, give them the stretch test.
- ☐ Spin all the pulleys to make sure they spin freely.
- ☐ Lubricate all the pulley shafts.
- ☐ Replace the belts back onto their proper pulleys.
- ☐ Lubricate all moving parts.
- ☐ Do not over lubricate. One drop goes a long way.
- ☐ Check the grease points for lubrication.

27

Checking and replacing motors

A VCR CONTAINS TWO TO FIVE SEPARATE DC MOTORS. DC STANDS FOR DIRECT current. This chapter covers the function of and how to identify and locate each motor. Refer back to chapter 25, "Getting into the undercarriage," before proceeding with this chapter.

Identifying dc motors

Look for a round cylinder. The cylinder is either brass or silver in color. You'll find two wires attached to the base and a shaft protruding out the front of it. There's either a gear or pulley mounted to this shaft. When the motor is activated, the gear or pulley drives the corresponding parts.

Cassette housing loading motors

Cassette housing loading motors are found only in frontload VCRs. This motor drives the cassette carriage and pulls the video cassette into the unit. When you push Eject, this motor drives the cassette carriage to push the video cassette back out. You will find this motor in one of three locations. The first location is on the right side of the cassette carriage. Its shaft is directly connected to the worm gear driving the carriage, as shown in Fig. 27-1. The second location is at the rear of the carriage near the right corner, with a belt and pulley driving the cassette carriage (Fig. 27-2). The third location is on the front of the cassette carriage, driving a worm gear. To retrieve this motor you must remove the cassette carriage. Refer to chapter 18, "Removing cassette carriages in frontload VCRs."

Loading motors

The loading motor controls the play, fast forward, rewind, and stop modes. It also is the vehicle the video tape uses to load or unload the video tape onto the video tape path. There are seven different locations for loading motors.

27-1   A worm-gear-driven cassette housing motor.

27-2   A belt-driven cassette housing motor.

The first location is horizontally set to the transport. The transport is the foundation all mechanical parts are connected to; it's the chassis of the complete carriage. In this model, the motor has a belt and a pulley attached to its shaft. Its shaft is always pointing to the rear of the unit. This type of motor is either on the right side of the transport (Fig. 27-3) or mounted the same way on the left side of the transport.

The second location is horizontally set to the transport. This type has two belts instead of one, as shown in Fig. 27-2. This type of loading motor is a combination motor, having two functions. It acts as a loading and cassette housing motor. There is only on location for this type of motor.

The third location is horizontally set as well and is on the right or left side of the under-carriage. This motor can have a worm gear attached to its shaft or a belt and pulley, as shown in Fig. 27-23.

The fourth location is set in a vertical position, as shown in Fig. 27-4. The motor is on top of the transport. Its belt and pulley are in the undercarriage. There is only one location for this type of motor.

To locate any other loading motor, place the unit on its side and open the undercarriage. Push Play and observe all the moving parts. Look for the motor driving all the gears in the undercarriage. The loading motor stops running after the tape is loaded onto the video tape path. All the other motors keep running. The loading motor also unloads the video tape, with all the gears operating in the opposite direction.

27-3 The location of a loading motor.

27-4 Clockwise from bottom left: the cassette housing motor, the loading motor, and the capstan motor.

Belt-driven capstan motors

There are two locations to find the belt-driven capstan motors. The first location is to the right of the video drum, as shown in Fig. 27-4. It's the largest vertically mounted motor. In some models, it has a flywheel on top of it. Its belts and pulley are found in the undercarriage. You'll find one to three drive belts on its pulley. This capstan motor drives the capstan shaft. The capstan shaft pulls the video tape through the video tape path. This motor drives the take-up spindle, which pulls the video tape back into the video cassette when in play. This motor also drives the fast forward and the supply spindle to rewind the video tape. All these functions use drive belts coming off the capstan motor's pulley.

Study Fig. 27-4. It shows a combination of three different motors mounted to the right of the video drum. From left to right is the loading motor, the cassette housing motor, and the capstan motor. The cassette housing motor is in this group of motors only in frontload models. In some models, the capstan motor is by itself in the same location.

The second location for the capstan motor is in the undercarriage, as shown in Fig. 27-5. This motor is mounted vertically to the transport, with its belt and pulley protruding through the top of the transport. The main drive pulley to the rear of the capstan motor is being driven by the capstan motor. The belts on the main drive pulley drive the rest of unit. This type of model only has two motors.

Direct drive capstan motors

The direct drive capstan motor is directly under the capstan shaft in the undercarriage. It is found only in newer models. The actual base, or flywheel, of the capstan shaft is part of the motor. This flywheel is the largest wheel in the undercarriage.

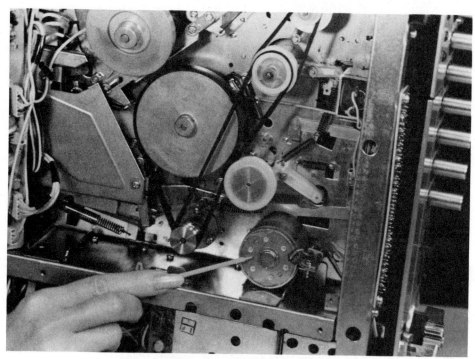

27-5 A location for a belt-driven capstan motor.

To identify a direct drive capstan motor, locate the circuit board protruding out from underneath the large flywheel, as shown in Fig. 27-6. In some models, the circuit board completely surrounds the flywheel (Fig. 27-7). The combination of the circuit board and flywheel makes the motor.

In some models, the direct drive motor has a reel drive belt attached to a small pulley in the center of the flywheel, as shown in Fig. 27-8. This belt drives the take-up spindle, fast forward, and rewind. Don't get this system confused with a belt-driven system. There's no other motor driving the flywheel; the flywheel is the motor itself. If you have a belt driven system, the belt is around the outside edge of the flywheel, as shown in Fig. 27-5. The belt means this system isn't the capstan motor, because the flywheel is driven by this belt. If a direct drive capstan motor isn't functioning, you have a circuit problem. You'll have to take the unit to a local service center for repairs.

Reel motors

There are two types of reel motors. The first type is located right under the take-up and supply spindles in the undercarriage, as shown in Fig. 27-9. The motor on the right is called the supply reel motor and is used for rewind. The motor on the left is called the take-up motor and is used for fast forward. When in play, this motor also pulls the video tape back into the video cassette. In Fig. 27-9, the positioning of the motors is shown, looking from the bottom of the unit.

Some models have only one motor, rather than two. This motor is located under and between the two spindles in the undercarriage, as shown in Fig. 27-10. The motor drives all three functions (fast forward, rewind, and take-up) with the aid of an idler wheel.

27-6   A circuit board protruding out under the flywheel.

27-7   A circuit board completely surrounding the flywheel.

27-8   A reel belt attached to the center of the flywheel.

27-9   The take-up reel motor (left) and the supply reel motor (right).

27-10 A reel motor.

Drum motors

The drum motor is the last type of motor found in a VCR. The drum motor is located directly under the video drum. The drum motor rarely goes bad. If it does, take the unit, with the make and model number, into a local service center for repairs. One cause to look for is a piece of video tape caught between the upper and lower video drums.

Checking motors for dead spots

If a motor isn't starting up, a coil in it might have opened. An open coil is a dead spot. A coil is wire wrapped around metal in a motor. The metal turns into an electrical magnet when electricity flows through the coil. Locate the spindle, pulley, or worm gear directly connected to the shaft of the motor. Spin it in either direction immediately after pushing the corresponding mode button. If the motor starts operating on its own, it has a dead spot. If a motor has a dead spot it has to be replaced.

Which mode button activates the correct motor

Motor	Button
Loading	Play or Stop
Capstan	Play, Fast Forward, or Rewind
Supply reel	Rewind
Take-up reel	Fast Forward

Drum   Play

Cassette housing   Eject (if a video cassette is in the unit)
                         * Insert a video cassette to activate the cassette housing motor when it is empty.

Locating frozen motors

If you can't spin the pulley when pushing the correct mode button, remove the drive belt from the pulley attached to the motor shaft. Then, push the correct mode button to see if the motor runs without a belt. If it does, check all the connecting moving parts this belt drives. A bent bracket or lever or a foreign object might be jamming the unit. If the motor doesn't run after removing the belt off the pulley, then check for bad bearings.

If a motor has a worm gear attached to its shaft and you can't turn the gear while pushing the correct mode button, refer to the section in chapter 9 called "Second reason" under the subhead "Reasons why the cassette holder jams." If you find no foreign object, remove the motor, leaving the leads to the motor attached. Then, push the correct mode button to see if the motor runs. If it doesn't run, check for bad bearings.

Finding bad bearings

If the shaft of the motor turns slowly or has a loud squeal, follow these next steps. Turn the power off and remove the motor's drive belt. If a worm gear is attached to the motor, remove the motor. Spin the pulley or worm gear on the shaft of the motor. Check it in both directions to see if it turns freely. If one bearing is bad, the pulley or gear will be hard to turn at that spot. If all the bearings are bad, it will be completely hard to turn or won't turn at all. If this is the case, the bearings are bad and motor has to be replaced.

Checking direct drive reel motors

To check direct drive reel motors without getting into the undercarriage, look between the two spindles. This type of system has no pulleys, wheels, or gears between the two spindles. All you'll find is a brake shoe assembly. The two spindles are directly connected by a shaft to the reel motors, as shown in Fig. 27-11.

Reel motors control the fast forward and rewind functions. Start by referring to chapter 15, "Deactivating tape sensors," and chapter 17, "Making a frontload VCR run without a video cassette." Push Rewind. Check the supply spindle on the left to see if it's turning. If not, give the spindle a quick spin counterclockwise. If the spindle operates on its own, there's a dead spot in the motor and the motor has to be replaced. If the spindle doesn't operate on its own, push Stop and look to the right side of the spindle for a brake shoe. Pull the brake shoe away from the spindle, as shown in Fig. 27-12. While holding back the brake shoe, turn the supply spindle to see if it spins freely. If the spindle is hard to turn, the bearings are bad and the motor has to be replaced. If neither of these instances is the case, the problem is the circuitry. If you have a circuit problem, take your VCR to a local service center for repairs.

Locate the take-up spindle on the right. Push Fast Forward. If the take-up spindle doesn't turn, give it a quick spin clockwise. If it starts turning, the motor has a dead spot and has to be replaced. Check for bad bearings by following the same procedure explained in the section in this chapter on "Finding bad bearings."

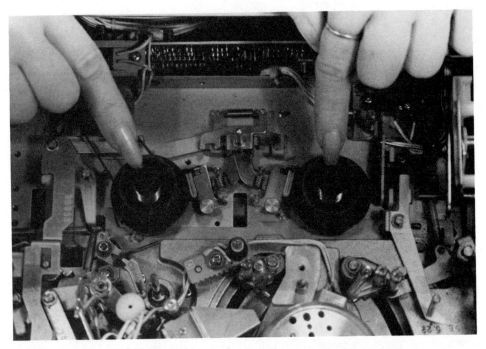

27-11   The supply spindle (left) and the take-up spindle (right).

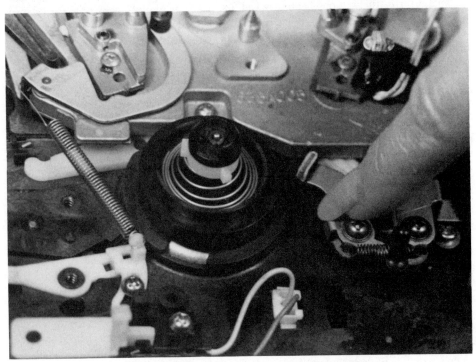

27-12   Pulling the brake shoe away from the spindle.

Removing spindles attached to a motor

There are two mounting systems used for spindles. One system has two lock nuts at the base of the spindle. The other system has an O-ring at the top of the spindle. Look for an O-ring on the top of the spindle to determine which type of system you have.

First type

For spindle systems using base lock nuts, measure the height from the transport to the top edge of the base of the spindle (Fig. 27-13). *This measure is very important.* Write down the measured distance for reference when remounting it. There are two lock nuts holding the spindle onto the motor shaft. These nuts are located directly under the spindle. Take a 1.5mm allen wrench and loosen both lock nuts by turning them counterclockwise, as shown in Fig. 27-14. Then, pull the spindle straight up and off.

After removing the spindle, you will be able to see what the two allen set screws look like and where they're located on the base of the spindle, as shown in Fig. 27-15.

To remount the spindle on the motor shaft, insert a allen wrench into one of the lock nuts and leave it there. Then, slide the spindle down the shaft. As you slide the spindle down

27-13 Measuring the distance from the transport to the spindle.

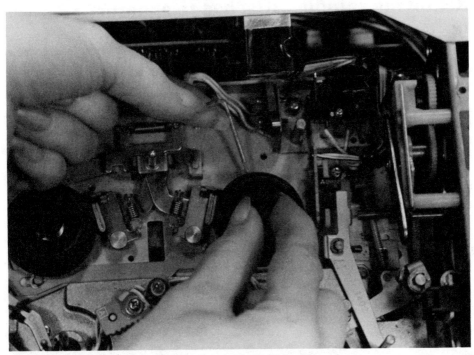

27-14　Loosening the lock nuts.

the shaft, locate the brake shoe and pull it away from the spindle. Refer to Fig. 27-12. The spindle must pass by the brake shoe. Next, tighten the lock nut holding the allen wrench. Tighten the nut until it's just snug. Leave it loose enough to move the spindle up and down the shaft. Now, remeasure the distance between the transport and the top of the base of the spindle, as previously shown in Fig. 27-13. Position the spindle to the correct measurement and tighten down the lock nut. Now, remove the allen wrench, insert it into the other lock nut, and tighten it. It's important that the measured distance be exact. If the measurement is off, it causes the reel inside the video cassette to rub on the inside of its cover.

For removing and remounting of the take-up spindle, follow the same procedure as previously explained. The exception is that the take-up spindle has a tape counter belt attached to it. Keep the belt tight when pulling it off the spindle. Slip the belt over any bracket it will reach (Fig. 27-16), so the other end of the belt doesn't come off the pulley on the tape counter. In most models, the tape counter pulley is hard to reach. If the belt comes off the pulley, you might have to remove the front cover to remount it. When replacing the belt onto the take-up spindle be sure the belt is in the groove.

Second type

The second type of system has an O-ring located on the top of the spindle. First, remove the O-ring, then pull the spindle straight up and off. After removing the spindle, you'll see a brass base connected to the motor shaft on which the spindle was sitting. Measure the height from the top of the dc motor to the top of this brass base. (See Fig. 27-19) Remember to write this measurement down, you'll need it to realign the spindle. Remove the brass base by

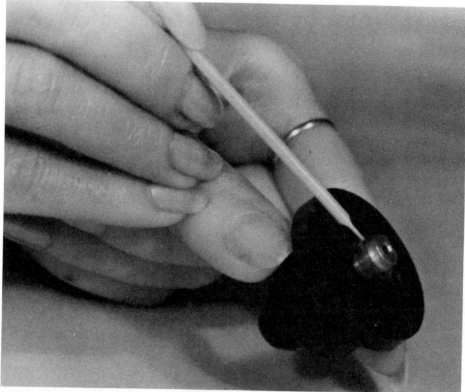

27-15 Location of the lock nuts on the spindle.

removing both allen lock nuts located on each side. Pull the brass base straight up and off the shaft.

When remounting the brass base onto the motor shaft, reposition the base to its correctly measured height as previously shown. Then, tighten down the lock nuts. Look at the bottom of the spindle. You can see that it's the same shape as the outside of the brass base. Line up these two patterns when remounting the spindle. If they are not aligned properly, the spindle won't slide down all the way. Remember to push aside the brake shoe as previously shown. The brake shoe can keep the spindle from sliding down all the way. Place the O-ring over the top of the shaft. With the aide of a small flathead screwdriver, push the O-ring down securely in the groove near the top of the shaft, as shown in Fig. 27-18. If not properly aligned, the O-ring won't fit or reach the groove.

Removing O-rings

Place a small flathead screwdriver under one side of the splits in the O-ring and pry that side up, as shown in Fig. 27-17. Next, place a finger on top of the O-ring where the screwdriver is and leave it there. Pull the O-ring up and off, with the aid of the screwdriver. Using the screwdriver will keep the O-ring from flying across the room. Be sure to put the ring in a safe place for remounting.

27-16 Stretching the counter belt over to a bracket.

27-17 Removing an O-ring.

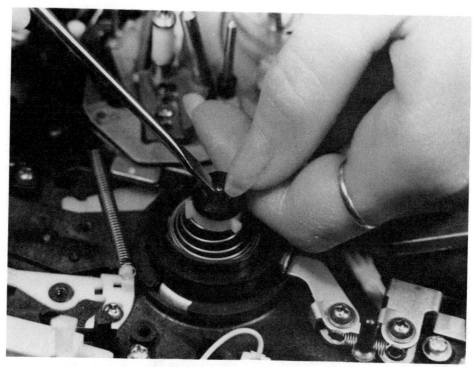

27-18   Replacing an O-ring.

Replace the O-ring. Push one side of it over the top of the shaft and down into its groove, with the aide of a screwdriver. Then, place your finger on the section you just put in the groove and use a screwdriver to push the rest of the O-ring down into the groove, as shown in Fig. 27-18.

Removing a reel motor

Unplug the VCR and place the unit on its side. Remove the bottom cover plate and open up the circuit board to the undercarriage. Refer to chapter 25, "Getting into the undercarriage." You now have a clear view of the take-up and supply motors. Refer to Fig. 27-9 for the location of these motors. The base of each motor has two separate wires attached. Follow the wire coming from the reel motor to a plug, as shown in Fig. 27-20. Unplug the wire.

Some models have no plug at the end of the wires connected to the reel motor. Instead, the wires are soldered onto terminals on the back of the motor. You'll find one small wire soldered to each terminal. Each wire will be a different color. There's a plus and a minus sign beside each terminal (Fig. 27-21). Draw a diagram of what colored wire goes to which marked terminal. This diagram will keep you from confusing the wires when reassembling. To detach the wires, cut them off close to each terminal with a pair of wire cutters. These wires have to be resoldered when you replace the motor.

Now, you're ready to remove the motor. Leave the unit on its side. Remove the two or three Phillips head screws located on top of the transport under the spindle you removed previously. At the same time, you need to hold onto the motor from the undercarriage side

27-19   Measuring from the top of the motor to the top of the brass base.

27-20   Unplugging the motor.

27-21   A plus and minus sign beside each terminal.

to keep the motor from falling out when you remove the last screw (Fig. 27-22). Pull the motor straight out from the undercarriage side. Take this motor with you to a local electronic supply store. To install the new motor, simply reverse the process.

Removing loading motors

Loading motors have three common mountings. One mounting uses clips. To remove the motor, spread the clips and pry the motor out. These motors are mounted horizontally, as shown in Fig. 27-2. The second type of mounting is a bracket over the motor with a screw on one side. Pull off the belt on the left and remove the screw (Fig. 27-23). Pull that side up and slide the motor out. To remount it, slide the sleeve on the side opposite to the transport, push the motor back into place, and remount the screw. To replace the belt on the right in this type of model, remove the motor first. The third type has three mounting screws, as shown in Fig. 27-24. Remove the belt and the three screws. Lift the motor straight up and out. Although it appears you can't replace the belt without removing the motor, you can in this type of model. For any motors not covered in this section, proceed to the last section of this chapter.

Soldering wires to terminals

After installing the new motor, you'll need to solder the wires back to the terminals on the back of the motor. Take wire cutters and strip off about ¼ inch of the insulation covering the

27-22   Removing a reel motor.

27-23   Removing the belt and mounting screw to remove the motor.

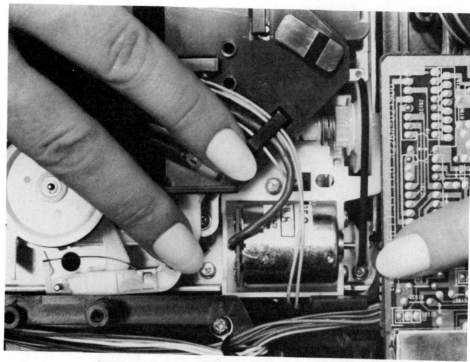

27-24 Removing the three screws to remove the motor.

wires. Put the right colored wire through the correct terminal hole and wrap the wire around the terminal. Take the soldering iron and coat the tip with solder. Flip off any excess solder. Touch the tip of the soldering iron and the solder directly to the terminal to form a bead of solder on the terminal, as shown in Fig. 27-25. Remove the solder, leaving the soldering iron in place for a few seconds to bond the wire to the terminal. Repeat this procedure for the other terminal.

Replacing a reel motor using a gear

Another type of direct drive system has a gear directly connected to the reel motor shaft. This gear drives the take-up spindle. Two Phillips head screws hold the motor in place, one on each side of the gear (Fig. 27-26). Take a flathead screwdriver and pry the gear off the motor shaft. Put the gear aside. You'll need it for reassembling. Follow the same procedure as previously explained to replace the reel motor.

After replacing the new motor, turn the unit around and replace the gear you previously removed. Slide the gear down the shaft until the teeth on the gear are perfectly aligned to the teeth on the take-up spindle, as shown in Fig. 27-26.

Replacing all other dc motors

All motors have either a worm gear, a drive wheel, a spindle, or a belt and pulley attached to their shaft. All worm gears slip right off. All drive wheels are attached with a locking nut.

27-25 Soldering a wire onto a terminal.

27-26 Aligning the teeth of the gear to the take-up spindle.

Most pulleys pry off the shaft. If the piece doesn't come off easily, look for lock nuts located on the base of the gear, wheel, or pulley. Take a 1.5mm allen wrench and loosen these lock nuts to remove the part.

Locate the wires attached to the back of the motor. Follow the wires to a plug and unplug it. If there's no plug, cut the wires off and resolder them, as previously explained. Refer to Figs. 27-22 and 27-26.

All motors are mounted by one to three Phillips head mounting screws. If the motor is mounted to the bottom of the transport in the undercarriage (Fig. 27-5), the mounting screws come down through the top of the transport to the motor. If the motor is on top of the transport (Fig. 27-4), the mounting screws are in the undercarriage. All other motors are horizontally mounted to a mounting bracket. In Fig. 27-4, the cassette housing motor is mounted to a bracket. The mounting screws are on one side of the bracket, and the motor is on the other. In some models, you'll have to remove the bracket to remove the motor. Locate the screws on the bracket and remove them. Now, the bracket will come off with the motor.

Review

- [ ] If the motor doesn't run, push the correct mode button and try spinning it to get it started to check for dead spots.
- [ ] Make sure you push the right mode button to activate the right motor.
- [ ] Check the motor for bad bearings.
- [ ] If the motor shaft doesn't move at all, disconnect the shaft from any belt or gear. If it starts running, then check for any foreign object or bent bracket that could be jamming it.
- [ ] If the motor is attached to a spindle, be sure to measure the height before removing the spindle.
- [ ] When replacing a spindle, be sure to remeasure it for proper height.
- [ ] All motors are mounted by two or three Phillips head screws.
- [ ] Remember that all motors are mounted on the opposite side of the transport or bracket from where the motor is located.
- [ ] When replacing the dc motor, be sure to connect the correct wire to the correct terminal on the motor.

28
Take-up spindle problems

THE TAKE-UP SPINDLE ATTACHES TO THE TAKE-UP REEL INSIDE THE VIDEO cassette when the cassette is inserted. The function of the take-up spindle is to pull the video tape back into the cassette after its gone through the video tape path. The other function of the take-up spindle is to fast forward the video tape in the cassette. For a better understanding of this function, read chapter 29.

VCRs stopping automatically

You're viewing a video tape in playback, there's good picture and sound, and then the unit automatically unloads the tape back into the cassette after running for only 30 to 60 seconds. If you try to push Play again, the same thing happens. You need to prepare the unit before you can make repairs to the unit.

First, remove the main top cover. (Refer to chapter 2, "Getting inside the VCR.") Place a piece of black electrical tape over each tape sensor. If you have a frontload unit, tape each tape sensor hole. (Refer to chapter 15, "Deactivating tape sensors.") Insert a video cassette and push Play. Look directly behind the pinch roller and capstan shaft. You'll see the video tape bundling up, as shown in Fig. 28-1. This continues for approximately 30 to 60 seconds, then the unit automatically stops and pulls the video tape back into the cassette. (I've removed the top cover on the video cassette to give you a better view.) Push Play again. This time watch the take-up reel inside the video cassette (Fig. 28-2). Notice the take-up reel isn't turning, or is turning sporadically. You'll need to look through the window on top of the video cassette to see the take-up reel.

The unit might run up to 30 minutes before this problem exist. The reason is that the take-up reel pulls in more video tape, which weighs down the reel. The weight and drag of the video tape causes the take-up spindle to stop turning.

28-1 The video tape bundling behind the pinch roller.

28-2 The take-up reel.

Take-up spindle systems

There are several types of spindle systems found in the various models of VCRs. Each type has its own set of instructions for correcting problems with the spindle. Read the following sections and follow the instructions that pertain to the model you have.

First type

If you have a topload model, refer to chapter 16, "Removing cassette carriages in topload VCRs." Push Play and locate the small drive wheel driving the take-up spindle, as shown in Fig. 28-3. The take-up spindle and the drive wheel should be turning. When the video cassette was in the unit the spindle stopped. The cause was drag from the weight of the video tape.

Now, I'll explain how the drive wheel comes into contact with the take-up spindle. Push Stop, then push Play. Watch the small drive wheel come over to the take-up spindle and drive the spindle. Place your finger on the take-up spindle to stop it from turning (Fig. 28-4). See if the drive wheel is spinning or has stopped. If it's still spinning, the spindle and the drive wheel have a buildup of residue. This residue causes the wheels to be slippery and has to be removed. If the drive wheel stops while you're holding the spindle, the problem is in the undercarriage. Undercarriage problems are covered in this chapter. Stop the unit and clean the parts.

Around the outside of either the take-up spindle or the drive wheel is a rubber tire. A rubber tire is placed between two wheels to cause friction. Which wheel the tire is on varies from manufacturer to manufacturer. In Fig. 28-4, the rubber tire is around the spindle.

28-3   The take-up drive reel.

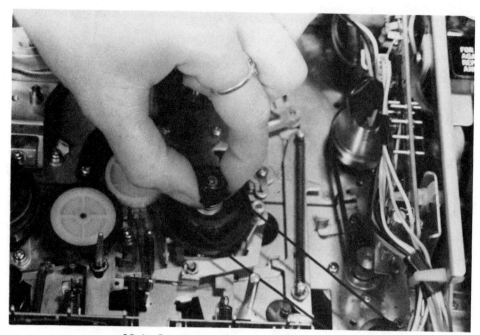

28-4 Stopping the take-up spindle from turning.

Cleaning rubber tires   Take a piece of 400 to 600 fine grain sandpaper and fold it in half. Sand the tire up and down as you slowly turn the spindle, as shown in Fig. 28-5. Sand all the way around the spindle. If the tire is shiney or it has cracks or a notch in the rubber, you must replace the tire. Refer to the section in chapter 29 on "Replacing tires."

Cut a 6 inch piece of masking tape and make a ring out of it with the sticky side out. Place your hand inside the ring, as shown in Fig. 28-6. Turn the spindle with one hand hold place the masking tape up against the rubber on the spindle, as shown in Fig. 28-7. Watch the masking tape as you spin the spindle. As the black residue builds up on the tape reposition it to a clean spot. Recheck the spindle for white spots left on the rubber tire, sometimes the white adhesive from the masking tape comes off causing white spots. While holding the spindle still, stick a new piece of masking tape over each white spot and pull the tape back off. You might have to repeat this procedure several times to remove the spots.

Cleaning wheels and spindles made of plastic or metal   To clean wheels made of plastic or metal, saturate a chamois stick with head cleaner or alcohol. Rub the chamois stick in an up and down motion as you go around the wheel or spindle (Fig. 28-8). Remove the residue with the masking tape method as previously explained. The exception is that you dab the wheel with the tape as you work around it. Check for white spots after using masking tape and remove them as previously explained.

Removing a take-up clutch assembly   Remove the drive belt and the C- or E-ring off the pulley as shown in Fig. 28-9. Then detach the spring and pull the clutch assembly straight out, as shown in Fig. 28-10. Take the old part and the make and model number of the unit into a local electronics supply center for replacement. To replace the assembly, simply reverse the process. Types of clutch assemblies differ in appearance. Replace all assemblies by using this procedure.

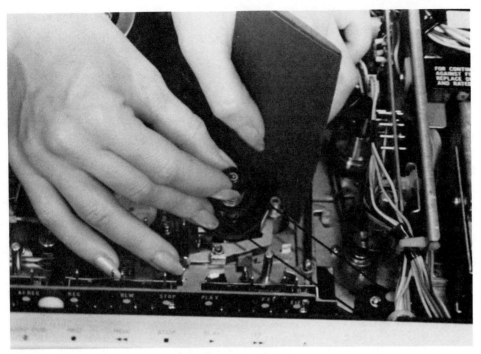

28-5 Sanding a rubber tire.

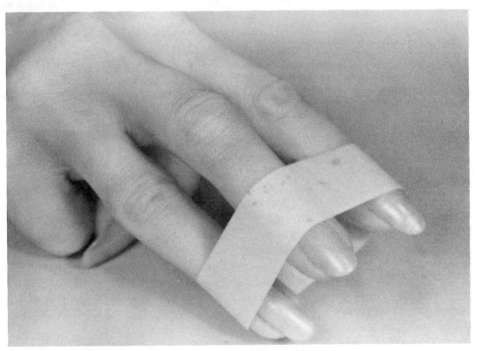

28-6 Making a ring from masking tape with the sticky side out.

28-7 Cleaning off the black residue from the tire.

28-8 Cleaning a metal drive wheel.

28-9 Removing the C-ring holding the clutch assembly.

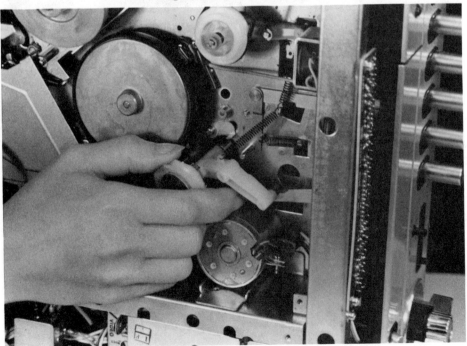

28-10 Removing the clutch assembly.

Second type

The second type of system uses a combination of a drive wheel, an idler wheel, and a take-up spindle. The idler wheel is to the right of the capstan shaft and behind the take-up spindle (Fig. 28-11). The drive wheel is at the base of the capstan shaft, as shown in Fig. 28-12. All drive wheels of this type barely stick out from under the base. When you push Play, the idler wheel moves over and makes contact simultaneously with the drive wheel at the base of the capstan shaft with the take-up spindle, as shown in Fig. 28-13.

Checking for slippage Prepare the VCR to run without a video cassette. Refer to chapter 15 for a topload unit and to chapter 17 for a frontload unit. Push Play and take hold of the take-up spindle to keep it from turning, as shown in Fig. 28-4. If the idler wheel stops turning but the drive wheel at the base of the capstan shaft is turning, the slippage is between the idler wheel and the drive wheel. On the other hand, if the idler wheel is turning, then the slippage is between the idler wheel and the take-up spindle. If both wheels stop turning, the problem is caused by a bad belt in the undercarriage. Check the belt attached to the flywheel.

All idler wheels have a rubber tire around them. The drive wheel and spindle are made of plastic. For cleaning the idler wheel, refer back to the section on "Cleaning rubber tires." For cleaning the spindle, refer back to the section on "Cleaning wheels made of plastic or metal." For cleaning the drive wheel at the base of the capstan shaft, you'll use a different cleaning technique.

Cleaning the capstan shaft base drive wheel Take a can of degreaser and spray off the wheel. Using a glass brush, remove any black residue left on the drive wheel and respray with degreaser. To rotate the wheel to clean the other side, place your finger on top of the flywheel on top the capstan motor and spin it in either direction (Fig. 4-6). This motion will cause the drive wheel to turn. To turn the drive wheel in models with no flywheel on top

28-11 The take-up idler wheel.

28-12 The take-up drive wheel.

the capstan motor, get into the undercarriage and turn the large flywheel at the base of the capstan shaft.

Checking idler wheels Sometimes after cleaning all three wheels, the take-up spindle still doesn't turn properly. If this situation occurs, you'll need to replace the tire on the idler wheel. First, remove the idler wheel assembly, then proceed to the section in chapter 29 on "Replacing Tires." If the idler wheel doesn't make good contact with the take-up spindle or the drive wheel, the idler wheel's clutch is slipping and the entire idler wheel assembly has to be replaced.

Removing a take-up idler wheel To replace the idler wheel assembly, remove the two brackets pointed at in Fig. 28-14 by removing the Phillips head screws at the base of each bracket. Remove the C- or E-ring holding on the pinch roller assembly (Fig. 28-15). For removing C- or E-rings, refer to chapter 7. Pick up the pinch roller and its bracket, pulling straight up and off. Remove the Phillips head screw on the left (Fig. 28-16). Lift the bracket straight up and off. Remove the C- or E-ring on the right that holds the idler wheel and bracket into place (Fig. 28-16). Lift the idler wheel with its bracket straight off and remove the spring attached to the bracket just above the idler wheel. Take the idler wheel and its bracket, along with the make and model number of the VCR, to a local electronics supply center for replacement. To remount the idler wheel, reverse the process.

Third type

The third type of system drives the take-up spindle, fast forward, and rewind. This system usually is found in frontload units. Because both systems work together, proceed to the

28-13 The point of contact for the take-up idler wheel and the drive wheel.

28-14 Removing the two brackets.

28-15 The C-ring holding on the pinch roller bracket.

sections in chapter 29 on "Idler arm and clutch assembly," "Checking the idler wheel," and "Replacing tires." Figure 29-7 illustrates this type of idler arm and clutch assembly.

Fourth type

In a few models, there's no idler or drive wheel on top of the transport, instead you'll find a cluster of gears between the spindles. The entire mechanism uses gears; there are no wheels to get dirty or cause slippage. Check the undercarriage for a broken belt. Refer to the first four sections of chapter 26. Check for a bad dc motor. Refer to chapter 27, "Checking and replacing dc motors."

Fifth type

The fifth type of system uses a direct drive system. You'll find no idler or drive wheel on top of the transport between the spindles. To check this type of system, refer back to the section in chapter 27 on "Reel Motors" under the subhead "Identifying dc motors." In some direct drive units, there's a small white gear beside the take-up spindle. Sometimes this gear will slip up or down the shaft, causing the spindle to stop turning. Refer to the section in chapter 29 on "VCRs using a direct drive system."

Spindle problems in the undercarriage

When holding the take-up spindle, the small drive wheel (Fig. 28-4) also stops turning. Place the VCR on its side and refer to chapter 25, "Getting into the undercarriage." Locate the

28-16 Removing the screw on the left and the C-ring on the right.

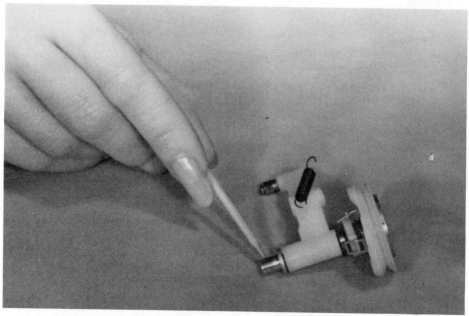

28-17 The clutch assembly and the drive wheel that drives the spindle.

belt and pulley to the small drive wheel. Push Play and manually stop the spindle from turning. If the pulley is turning, the clutch inside the assembly is worn or out of adjustment. If the pulley isn't turning, the belt has worn and is slipping. (I've removed the take-up clutch assembly in the figure for demonstration purposes only.)

At the top of the assembly is where the drive wheel comes into contact with the take-up spindle and drives it, as shown in Fig. 28-17. At the opposite end of the assembly is a pulley. This location is where the belt attaches. After a lot of use, the clutch in the take-up clutch assembly can slip. Look on the backside of the pulley where the belt attaches. On some assemblies you'll find a clutch pressure adjustment with three fingers and four notches on the back of the pulley (Fig. 28-18). To readjust the pressure on the clutch, place a small screwdriver in the slot, pull the spring assembly back, and turn the pulley counterclockwise until the fingers sit into the top notches, as shown in Fig. 28-19. This adjustment keeps the clutch from slipping.

If you have an assembly without a clutch adjustment or it's adjusted all the way and the pulley turns but not the wheel, then the entire assembly has to be replaced.

Review

☐ If the VCR stops after 60 seconds while in play, you might have a take-up spindle problem.

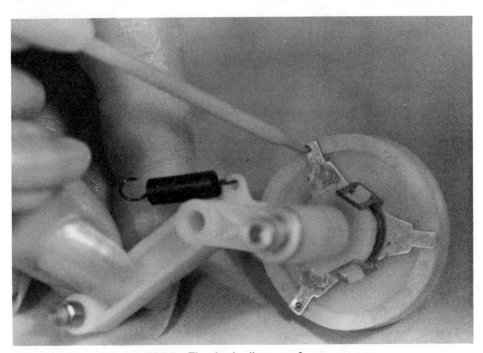

28-18   The clutch adjustment fingers.

☐ Check to see if any excess video tape is coming out from behind the pinch roller and capstan shaft.

☐ Check to see if the take-up spindle and all connecting wheels are turning. If they are not, check the undercarriage for a bad or broken belt or a bad dc motor.

☐ Check for slippage between each wheel and clean them, if needed.

☐ If a rubber tire is shiney, has cracks, or a notch in it, replace the tire.

☐ If the idler wheel doesn't make contact with the spindle, check for a bad clutch system and replace it.

☐ If you have a direct drive system, check the dc motor.

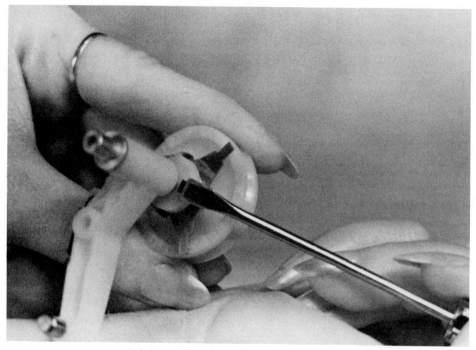

28-19   Adjusting the fingers on the clutch.

29

Fast forward and rewind problems

THERE ARE THREE PROBLEMS THAT COMMONLY OCCUR IN FAST FORWARD AND rewind systems. One system or the other malfunctions, the VCR runs slow, or the unit stops before rewinding the entire tape.

Preparation for repair

In topload units, remove the cassette carriage and make the unit run without a video cassette. Refer to chapter 16, "Removing cassette carriages in topload VCRs." (It also is possible to make this repair without removing the carriage. Refer to chapter 15, "Deactivating tape sensors." In this case, it'll be harder to see the moving parts and work on it.) For frontload units, refer to chapter 17, "Making a frontload VCR run without a video cassette." You will be able to see all the moving parts and where the problems exist. This preparation must be done before you can work on fast forward or rewind problems. In frontload units, push Eject to pop up the cassette holder for removing or repairing any parts.

The fast forward system and how it functions

There are three moving parts on top of the transport in older models. The combination of these three parts make up the fast forward system, as shown in Fig. 29-1. There's a belt and pulley in the undercarriage, which turn the drive wheel. The drive wheel drives the idler wheel. The idler wheel drives the spindle. The spindle drives the video cassette. Push Fast Forward and observe how the three wheels come together and drive the take-up spindle. You'll notice the idler wheel moves in the direction shown in Fig. 29-2, making contact with the drive wheel and the spindle simultaneously.

When you inserted a video cassette, you probably noticed that the video tape wasn't moving or was turning slowly. Since the unit now is running without a video cassette, the

29-1 From left to right: the drive wheel, the idler wheel, and the take-up spindle.

29-2 The direction the idler wheel moves to make contact with the drive wheel and spindle.

take-up spindle will operate correctly. The video cassette isn't in the unit to put drag on the take-up spindle and make the unit slow down or stop.

Checking the fast forward system

Keep the unit in the fast forward mode. Grab the take-up spindle and stop it from turning. Notice if the idler wheel and/or drive wheel has stopped turning. If all the wheels have stopped, the problem is in the undercarriage. You'll find a bad drive belt connected to the drive wheel. The belt is slipping on its pulley. In this case, refer to chapter 25, "Getting into the undercarriage," and to the first four sections of chapter 26.

If only the idler wheel stops turning, then you have slippage between the idler and drive wheels, as shown in Fig. 29-3. If both the drive wheel and the idler wheel continue turning while you're holding the take-up spindle, the problem is between the idler wheel and the take-up spindle, as shown in 29-4. Clean each wheel. (Refer to the sections in chapter 28 on "Cleaning rubber tires" and "Cleaning wheels and spindles made of plastic or metal." If the cleaning didn't repair the problem, proceed to the section in this chapter on "Replacing rubber tires."

When holding the take-up spindle to keep it from turning, the unit will stop periodically. There's nothing wrong. A sensor signals the unit to stop when the take-up spindle isn't turning. When the sensor stops the unit, push Fast Forward again.

29-3   The location of the slippage.

29-4 The location of the slippage.

Checking the rewind system

There are two wheels involved in the rewind system: the drive wheel and the supply spindle. Push Rewind and observe how the drive wheel moves to the supply reel in the direction shown in Fig. 29-5. To check this system take hold of the supply spindle and stop it from turning. The unit will periodically stop as previously explained, push Rewind again. If the drive wheel stops turning at the same time, the problem is a bad belt in the undercarriage. If it continues to turn, the problem is between the drive wheel and the supply spindle, as indicated in Fig. 29-6. You'll need to clean the spindle. Refer to the section in chapter 28 on "Cleaning spindles made of plastic or metal." Then proceed to the section in this chapter on "Replacing tires."

Idler arm and clutch assembly

In the newer models, there is a long arm with an idler wheel attached to it. This part is called an idler arm, and clutch assembly (Fig. 29-7). When in fast forward or play, this assembly flops over to the take-up spindle and drives it. When in rewind, the assembly flops over to the supply spindle and drives it. Notice that there is a drive wheel right at the swivel point of the arm (Fig. 29-8). This drive wheel drives the idler wheel, which in turn drives the spindles. The drive wheel is powered by a belt and pulley in the undercarriage.

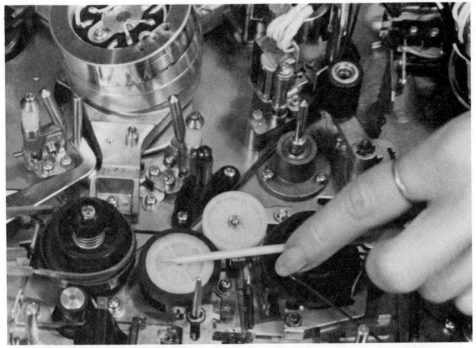

29-5 The direction the drive wheel moves to make contact with the supply spindle.

29-6 The location of the slippage.

29-7 The idler arm assembly.

29-8 The drive wheel on the clutch assembly.

Checking the idler wheel

To check this type of system, place the unit into either fast forward or rewind, depending on what type of problem you have. Take hold of the proper spindle, which is the spindle that is turning, and stop it. While holding the spindle, check to see if the idler wheel stops turning. If it continues to turn, then you have slippage between the spindle you are holding and the idler wheel. If the idler wheel stops turning, then there's slippage between the idler wheel and drive wheel. If the drive wheel and idler wheel both stop turning, then the problem is a bad belt in the undercarriage.

The majority of the time, its the idler wheel causing the slippage. Clean each spindle and the drive wheel. Refer to the section in chapter 28 on "Cleaning wheels and spindles made of plastic or metal." Then, replace the tire on the idler wheel. Refer to the section in this chapter on "Replacing tires."

Some frontload models use a gear system instead of a drive wheel. This gear drives the idler wheel. The idler wheel or idler arm flops from spindle to spindle. The mounting bracket covers most of the idler wheel. Check this system as previously explained, except notice that the idler wheel is hard to reach. To clean the idler wheel, place a small flathead screwdriver on one side of the wheel to steady it. Use a glass brush to clean the other side, as shown in Fig. 29-9. If, after cleaning the idler wheel and the drive wheel, the unit still doesn't work correctly, then you'll need to relace the tire on the idler wheel.

In another type of system, a shaft is sticking through a curved slot on the transport, as shown in Fig. 29-10. The shaft is attached to the idler wheel. Check this system for slippage as previously explained. Get into the undercarriage to clean or replace the idler wheel. Proceed to the appropriate section in this chapter for removing idler arms and clutch assemblies.

Another type of system has a large idler wheel between both spindles, as shown in Fig. 29-11. No idler arm is attached to the idler wheel. In fast forward or rewind the idler wheel itself slides over to the proper spindle. The drive wheel that drives the idler wheel is in front of the idler wheel under a metal bracket. Stop the proper spindle to check for slippage. Then refer to the section in chapter 28 on "Cleaning wheels and spindles made of plastic or metal." Remove the E-ring off the idler wheel and pull it off. Then, replace the rubber tire on the idler wheel. Refer to the section in this chapter on "Replacing tires." If the idler wheel doesn't make complete contact with either spindle, the idler wheel has a bad clutch assembly and has to be replaced.

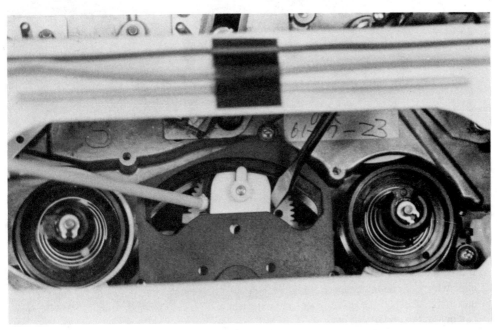

29-9　Using a glass brush to clean the idler wheel.

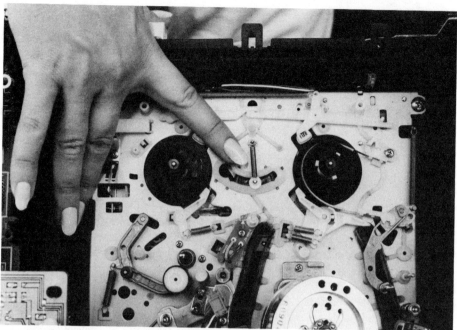

29-10   An idler wheel shaft protruding through a curved slot between the spindles.

29-11   The drive wheel connected to the idler wheel.

Replacing tires

There are two different ways to replace tires in VCRs. The majority use the first method. Read both sections, however, to determine which of the two ways is applicable to your unit. Follow the instructions given in that section.

First way

In 90% of all VCRs, the tire is mounted to the idler wheel. If the idler wheel is exposed on the top of the transport, replace the tire by removing any springs attached to the shaft of the idler wheel, as shown in Fig. 29-12. Take a cuticle remover and nail cleaner tool or a nut digger and insert it between underneath the plastic on the idler wheel and the tire. (See Fig. 29-13) Pry the tire out, being cautious not to break the plastic on the idler wheel. Take a second nail tool, place it next to the first, and continue prying the tire out from the idler wheel. Hold it in this position. You should have a gap between the idler wheel and the drive wheel. Now, remove the first nail tool, place it in the gap between the tire and the idler wheel, and hold it there. Take a pair of long-nosed pliers and grab the tire where it's sticking out. Place a firm grip on the tire. Stretch the tire away from the idler wheel and pull straight up and back over the wheel. Pull the idler wheel away from the drive wheel to remove the tire.

To replace the tire, place it in the groove of the idler wheel, halfway around. Turn the idler wheel around, so the part of the tire in the groove goes past the drive wheel, as shown in Fig. 29–14. Place the nail tool under the tire and on top of the idler wheel (Fig. 29-14).

29-12   The spring located on top of the idler wheel.

29-13 Inserting the tool between the wheel and tire.

29-14 Replacing a tire.

Pull straight up on the tool, and that section of tire will pop up over the plastic and into the groove. Keep a finger in the position shown in Fig. 29-14 to keep the tire from popping back out. Continue this process until the tire is correctly mounted.

Sometimes when you pop the tire onto the idler wheel, it won't seat correctly into the groove, making the tire look twisted on the idler wheel. Take the nail tool and push down on any places that aren't even. This step will place the tire completely, and correctly into the groove.

Another difficulty might be that the tire isn't all the way down into the groove, making the tire look lopsided. Take the tire and squeeze it until the tire is even all the way around.

Second way

In all other models, the idler wheel isn't exposed. You'll have to remove the complete clutch assembly to replace the tire. Refer to the section in this chapter on "Removing idler arms and clutch assemblies." To remove the tire, hold one side of the assembly while grabbing the tire, stretch it straight out, and fold it back over the wheel. For remounting, place one side of the tire in the groove. Stretch the other side over the wheel and into the groove. For replacing any other tires, such as on spindles or drive wheels, use the same procedure.

Checking an idler arm for a bad clutch assembly

In fast forward or rewind, the idler wheel doesn't come completely into contact with the proper spindle. The idler wheel is spinning, but the spindle that is supposed to turn with it isn't turning. This condition means the clutch inside the idler arm has gone bad. You'll need to replace the idler arm and clutch assembly. You also will need to replace the assembly if you put the unit into play and the idler wheel doesn't come into contact with the take-up spindle.

Another problem is when you push the fast forward or rewind and the idler arm won't flop over to its proper spindle. The idler arm doesn't move, and the idler wheel doesn't turn. The problem is in the undercarriage. You have a bad or broken belt or a bad dc motor. Refer to the first four sections of chapter 26 or to chapter 27, "Checking and replacing dc motors."

In idler assemblies using a gear instead of a wheel, the idler gear doesn't come all the way over to either spindle and make contact. The problem is that the unit has a bad clutch. You'll need to replace the whole assembly.

Removing idler arms and clutch assemblies

The steps that you will follow to remove the idler arm and clutch assembly in your VCR depends upon the model that you have. There are six different methods described in the following sections. Read each section and follow the proper instructions.

First type

Open the undercarriage and take the belt off the pulley on the bottom of the idler arm and clutch assembly. Remove the Phillips head screw on each side of the arm and clutch assembly, as shown in Fig. 29-15. Pull the entire assembly straight up and out (Fig. 29-16).

A similarly mounted assembly has two white gears on top of the transport. One gear is on each side of the idler arm (Fig. 29-17). Remove the same two screws. In a few models, the assembly lifts out, but in most models, the pulley in the undercarriage has to be removed

29-15   Locating the two Phillips head screws.

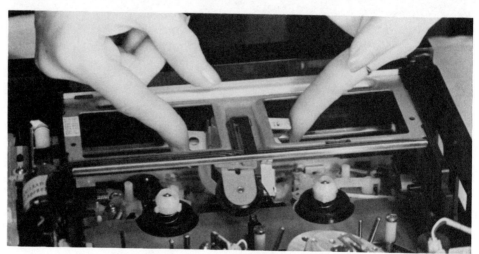

29-16   Removing the entire assembly.

first. Refer to chapter 25, "Getting into the undercarriage." Take off the belt and remove the C- or E-ring on top of the pulley. Pull the pulley straight off. Now, the tire can be replaced. If the clutch has gone bad, take the entire assembly, along with the make and model number of the VCR, to a local electronics supply center for replacement.

Second type

The second type of system is mounted with an O-ring on top of the shaft, right above the drive wheel on the arm and clutch assembly (Fig. 29-18). Remove the O-ring. Refer to

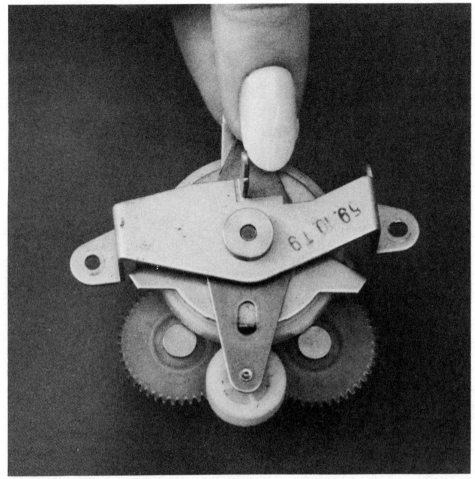

29-17   A white gear on each side of the idler arm.

the section in chapter 27 on "Removing O-rings." In some models, a spring is attached over the idler arm, as shown in Fig. 29-18. Remove this spring and slide the idler arm and clutch assembly straight up and off the shaft. In this type of system, don't remove the belt in the undercarriage.

Third type

This system has no E- or O-rings or screws mounting the assembly to the transport. What you'll find is a small white clip at the base of the arm, right above the drive wheel. Take this clip and pull it straight back to release the latch, as shown in Fig. 29-19. Pull the entire assembly straight up and off.

Fourth type

This type of system has no arm attached to the idler wheel (Fig. 29-20). It is attached to the transport by a bracket and the idler wheel slides back and forth from spindle to spindle. To

29-18 Clutch assembly with an O-ring.

29-19 Releasing the latch to remove the clutch assembly.

29-20 An idler clutch assembly with no arm attached.

replace the tire, remove the E-ring. Refer to the section in chapter 7 on "Removing E-rings" under the subhead "Finding the lubricating points." Slide the idler wheel from the drive wheel, pulling straight up and off. To remove this type of clutch assembly, remove the two screws on each side directly across the center of the idler wheel. Pull the assembly straight up and off. Figure 29-21 illustrates the clutch assembly and the mounting screw holes. In this type of system, the tire can be replaced without removing the clutch assembly.

Fifth type

This type of system has a large black mount over the top of the idler wheel assembly. Remove the belt off the idler wheel pulley in the undercarriage. Remove the three screws on top of the large black mount (Fig. 29-9). The screws are mounted in a triangle pattern. Pull the entire assembly straight up from the top of the transport. Now, you can replace the tire on the idler wheel.

Sixth type

In this type of system, the shaft of the idler wheel protrudes through the top of the transport between the two spindles, as shown in Fig. 29-10. Remove the take-up spindle by removing its O-ring. (There's a small washer under the spindle. Don't lose it.) Remove the small spring attached to the center of the shaft protruding through the transport. In the undercarriage, remove the drive belt attached to the pulley that drives the idler wheel. Remove the spring that is attached to the black cover and connected to a long white arm on the other end, as shown in Fig. 29-22. Wires run across the black cover through clips. Loosen the wires and

29-21 A removed clutch assembly and the location of the mounting screw holes.

29-22 The spring on the black cover.

29-23 Locating the screw and the two black clips.

pull them out of the clips. Next, remove the black plastic cover, which is held on with one screw and two clips (Fig. 29-23). Remove the two O-rings connected to the idler arm assembly and clutch assembly pulley, as shown in Fig. 29-24. Some models have the same assembly except that they have no black cover. Follow the same procedures, omitting instructions about the black cover.

First, rotate the clutch assembly arm to the left to the molded indentation in the transport. Take hold of the idler wheel and pull the assembly off. Some models have no molded indentation on the transport. In these models, place pressure with a flathead screwdriver on the brass center of the pulley. Then, take hold of the pulley and pull it straight up. The pulley pops right off the shaft. On a shaft underneath the pulley is a sandwich made up of a washer, a gear, and another washer. Take hold of the remaining assembly and pull it off. (The arm bends, but don't worry, it won't break.) Slide the idler wheel to the left to the circle at the end of the slot. Then, pull off the idler arm and idler wheel. Now, you can replace the tire.

To remount this assembly, reverse the process. If you hear a grinding noise in fast forward or rewind after reassembling the unit, one of the washers is incorrectly placed on the shaft of the clutch pulley.

VCRs using direct drive systems

The last system is a direct drive system. The motors are directly connected to the supply and take-up spindles. To determine if you have this type of system, refer back to the section in chapter 27 on "Checking direct drive reel motors." In some direct drive units, you'll find a

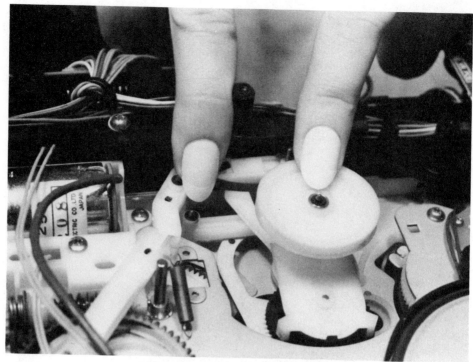

29-24 Locating the O-ring for removal.

white gear beside the take-up spindle. The white gear is directly connected to a reel motor that drives the take-up spindle.

When you put this system in fast forward, the take-up spindle doesn't turn and you hear a loud grinding noise. In play, you'll hear a clicking noise. In either case, after 60 seconds the unit stops. The problem is that the gear has worked itself up or down the shaft of the reel motor, as shown in Fig. 29-25. The teeth on the gear become out of alignment with the teeth on the take-up spindle. To correct this problem, place a flathead screwdriver under the gear and pry the gear up the shaft until the teeth on the gears are realigned.

A trade tip for fast forward and rewind problems

In some models, when you push Fast Forward or Rewind, the mode begins. Then, however, you hear a loud click and the unit stops. If you try again, the same problem occurs. To repair this problem, open the undercarriage. Then, look for a white L-shaped bracket with a screw inserted in the front portion of the transport, as shown in Fig. 29-26. The problem was caused by a small square rubber pad falling off this bracket. The location of the pad is as shown in Fig. 29-26. When the pad is missing, the linkage in the undercarriage is thrown out of synchronization. Locate the missing pad in the undercarriage and glue it back into place.

Review

☐ On topload units, remove the cassette carriage and prepare the unit to operate without a video cassette.

☐ On frontload units, drop the carriage down and prepare it to run without a video cassette.

☐ If fast forward, or rewind runs slow or stops, check for slippage between the drive wheel, idler wheel, and each spindle.

☐ Clean each spindle and drive wheel. Replace the tire on the idler wheel if necessary.

☐ If fast forward or rewind doesn't work at all, look at the clutch system to check for positive contact against the spindles.

☐ If the drive wheel doesn't turn, check the undercarriage for a broken belt or a bad dc motor.

29-25 The white gear connected to the take-up spindle.

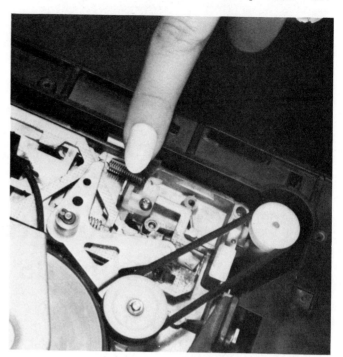

29-26 The rubber pad on the white bracket.

30

Tape counter problems

A TAPE COUNTER MEASURES THE NUMBER OF FEET OF VIDEO TAPE THAT have traveled through the video tape path. The counter gives you a reference point on the video tape. There are two types of tape counters. The first type is an electronically-produced digital tape counter. This type rarely shows a problem. The second type is a mechanical tape counter. In a mechanical tape counter, the numbers rotate around. This chapter is on mechanical type counters and their malfunctions.

The first problem occurs while the unit is in playback. You hear a loud clicking sound, or if the unit is in fast forward or rewind, you hear a grinding noise. When you hear this clicking or grinding sound, look at the tape counter to see if the numbers are rotating. Sometimes the gears in the counter don't engage properly after you've reset the counter to zero.

To correct this problem, push the reset button all the way in with the corner of your fingernail, as shown in Fig. 30-1. Quickly flip your fingernail off the side of the button, so the button pops out. This pressure pops the gears inside the counter back into place. You might need to do this procedure two or three times to reset the gears.

The second problem also occurs while in playback. The unit is playing, then automatically stops. If you push Play again, the unit plays for a minute, then stops again. If you put the unit in rewind or fast forward, the unit runs for about 30 seconds, then stops. The reason is that there's a sensor incorporated in all tape counters. When the tape counter pulley doesn't turn, this sensor signals the unit to stop.

To repair this problem, look at the tape counter while the unit is in playback to see if the numbers are rotating. If not, remove the main top cover. (In some models, you'll need to remove the front cover.) Proceed to the first five sections of chapter 35. Locate the tape counter belt. One end is attached to the take-up spindle. The other end is attached to a small pulley at the bottom of the tape counter, as shown in Fig. 30-2. There are three reasons why the tape counter won't rotate: The belt is broken, has slipped off one of its pulleys, or has stretched out of shape, causing slippage. Just replace the belt.

I've found instances where a unit might run for 30 minutes to an hour, and then stop. The cause of this intermittent problem is that the belt on the take-up spindle isn't positioned properly in the groove. The belt is slipping on the spindle itself.

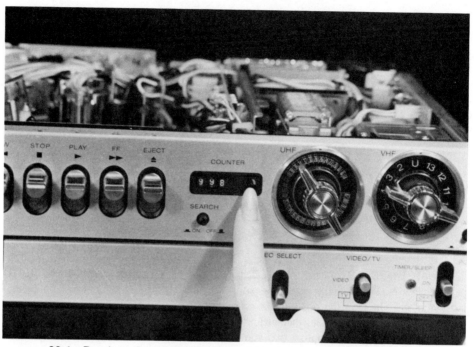

30-1   Popping a fingernail off the edge of the reset button to clear the gears.

30-2   The tape counter and its belt.

31

Capstan shaft problems

THE CAPSTAN SHAFT WORKS IN CONJUNCTION WITH THE PINCH ROLLER TO pull the video tape through the video tape path. A capstan shaft problem can cause the following symptoms. In play the VCR loads correctly. First, you have no picture at all, or the unit appears to be in pause, then after 30 to 60 seconds, the unit unloads and goes into the stop mode. Another symptom that you hear a loud squeal while the unit is in play, although you have a good picture. Each time you try to play the VCR, the same symptoms occur.

Detecting a capstan shaft problem

Remove the main top cover. (Refer to Chapter 2, "Getting inside a VCR.") Insert a video cassette and push Play. The unit immediately loads up the video tape onto the video tape path. Right after the unit loads up, the video tape stop moving and the reels inside the video cassette stop turning. After 60 seconds or so, the unit automatically unloads and the video tape pulls back into the video cassette.

Locate the capstan shaft and pinch roller. While the unit is in play, the video tape goes between these two parts, as shown in Fig. 31-1. Push play and watch the capstan shaft to see if it's turning. If the shaft is turning but the video tape isn't moving, the pinch roller isn't pressed up tightly against the capstan shaft. If this tension is missing, the shaft is unable to pull the video tape through the video tape path. The problem is that the pinch roller bracket is bent or there's a foreign object stopping the bracket from moving over to the capstan shaft. Refer to the section in chapter 10 called "Second reason" under the subhead "Reasons why a VCR will eat a video tape." Follow the instructions for removing the pinch roller bracket.

There are three different reasons why the capstan shaft would not be turning. A belt on the capstan shaft flywheel might be broken. Refer to the first four sections of chapter 26. A dc motor might have gone bad. Refer to chapter 27, "Checking and replacing dc motors."

31-1   The video tape between the capstan shaft and the pinch roller.

Finally, a capstan shaft might have frozen up. Refer to the section in this chapter on "Removing a frozen up capstan shaft."

Checking an older direct drive VCR

Some older direct drive VCRs have an oil well surrounding the base of the capstan shaft (Fig. 31-2). If the unit has an oil well, you should be able to freely spin the capstan shaft when the unit is in the stop mode. If the shaft won't turn or is difficult to turn, take a can of degreaser and spray degreaser into the slot around the top of the oil well, as shown in Fig. 31-3. The degreaser frees the bearings by removing any rust or dirt. After you're done spraying, the shaft should turn. The first few turns will be hard, but keep turning the shaft until it turns freely. After you've freed the shaft, lubricate the oil well with household oil until the felt is well saturated, but not overflowing. Then, turn the shaft until it spins freely. Be sure to wipe any excess oil off the capstan shaft. Insert a video cassette and push Play. Leave the unit in play for one or two hours, so the bearings lubricate properly.

Checking a belt-driven VCR

Belt-driven units are the most common type of VCR. These VCRs don't have an oil reserve at the base of the capstan shaft. You can't turn the shaft by hand.

Unplug the VCR, place the unit on its side, and remove the bottom cover plate. If necessary, open the main circuit board to reveal the undercarriage. Refer to chapter 25, "Getting into the undercarriage." You're looking for the largest wheel in the undercarriage. This wheel is called the a flywheel (Fig. 31-4). The flywheel is attached to the base of the capstan shaft. Remove the belt attached to the flywheel. In some models, there's a bracket

31-2 An oil well surrounding the base of the capstan shaft.

31-3 Spraying degreaser into the oil well.

31-4 Spinning the flywheel.

covering the flywheel, as shown in Fig. 31-5. Remove this bracket before removing the belt. Place your finger on the outer edge of the flywheel and spin it (Fig. 31-4). If it turns easily, the flywheel doesn't need to be lubricated. If the belt isn't broken you have a bad DC motor. Refer to chapter 27, "Checking and replacing dc motors." If the dc motor is okay, you have a circuit problem. Circuit problems have to be taken to a local service center to be repaired. If the flywheel is difficult or impossible to turn, you'll need to remove the capstan shaft for lubricating.

Removing the capstan shaft in belt-driven VCRs

Remove the screws at each end of the bracket at the base of the flywheel, as shown in Fig. 31-5. Remove the bracket and the drive belt. Take hold of the fly-wheel and work it out by twisting it back and forth as you pull on it. Continue until the flywheel is pulled out approximately ½ inch, then push the flywheel back in.

Look to the other side of the transport at the capstan shaft. You'll find a washer approximately ½ inch up the capstan shaft, as shown in Fig. 31-6. Remove the washer from the capstan shaft. (Be sure to put the washer in a safe place. The washer keeps the oil from coming up the capstan shaft and getting on the video tape.) Go to the undercarriage and pull the flywheel and capstan shaft all the way out. If you fail to follow this procedure, the washer on the capstan shaft will fall off into the transport and become lost.

Removing a frozen up capstan shaft

If the flywheel and the capstan shaft are completely frozen tight, remove the flywheel bracket and drive belt. Try to remove the flywheel assembly by twisting the flywheel back and forth,

31-5   The flywheel bracket and its mounting screws.

31-6   A washer halfway up the capstan shaft.

31-7 Tapping down the capstan shaft with a hammer and a block of wood.

pulling on it at the same time. If it doesn't move, place the unit right side up. Take a block of wood and a hammer and place the block on top of the capstan shaft (Fig. 31-7). Tap the block lightly with the hammer to loosen the capstan shaft for removal. Place the unit back on its side and remove the flywheel and capstan shaft as previously explained.

Cleaning and lubricating the capstan shaft

Take a paper towel and dip the corner into some cleaning solution. Place the capstan shaft onto the wet portion of the towel, as shown in Fig. 31-8. Twist it back and forth while pulling the capstan shaft out of the paper towel. This procedure removes all the dirt and residue from the capstan shaft. You might need to repeat this step several times. Take a can of degreaser, using the extended nozzle, and spray degreaser into the capstan shaft hole (Fig. 31-9). Take the glass brush and place it into the hole. Twist the brush back and forth as you push it all the way through. Pull the brush out, spray it with degreaser, and clean the brush off with a paper towel. Repeat this procedure until all the residue inside the hole is removed. After you've completed this process, spray out the hole to remove any fibers left by the glass brush.

Take a can of household oil and place the lubricant on the capstan shaft only. Start lubricating from the flywheel where it is attached to the capstan shaft, going up the shaft about an inch, as shown in Fig. 31-10. (Don't get any oil on the upper portion of the capstan shaft where the video tape makes contact.) Place the capstan shaft back into the capstan shaft hole. Push the flywheel in all the way. Leave the unit on its side and spin the flywheel. It should spin freely by itself for a minute or so. If it still turns a little hard or unevenly, repeat the procedure.

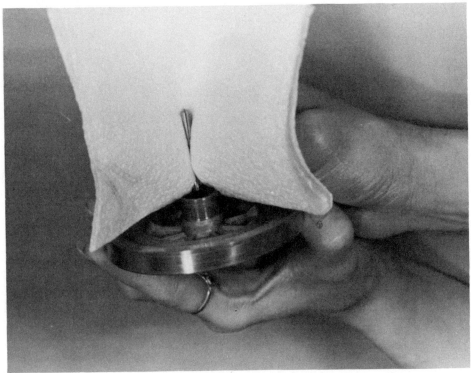

31-8   Cleaning the capstan shaft.

Remounting, greasing, and adjusting the capstan shaft

Before replacing the mounting bracket, look on the inside of the bracket. You'll find a plastic plateau in the center of the bracket. It usually is round in shape. On this plateau, you'll find old grease or no grease at all. Using a paper towel, remove the old grease, then apply a small amount of phono lube or tuner grease onto its center. Look at the center of the flywheel. You'll see a shaft protruding out about ¼ inch. This is the base of the flywheel. Take a paper towel and wipe off all the old grease on the end of the shaft. Replace it with new grease, as shown in Fig. 31-11.

After cleaning, lubricating, and greasing the flywheel and capstan shaft, it's time to re-mount the drive belt onto the outside of the flywheel. Place the mounting bracket back into the base of the flywheel and replace the two mounting screws.

Some models have an adjustment screw on the bottom of the bracket under the plateau (Fig. 31-12). Don't touch this adjustment screw, unless it is necessary. You can determine whether an adjustment is needed by moving the flywheel up and down. It should have no more than ¾ to 1 mm of play. If it's too tight, the capstan shaft won't turn freely. If it's too loose, the capstan shaft moves up and down, eating the video tape.

Remount the bottom cover plate and circuit board, if needed. Place the unit right side up. Replace the washer you set aside earlier. Slide it down the capstan shaft to the base of

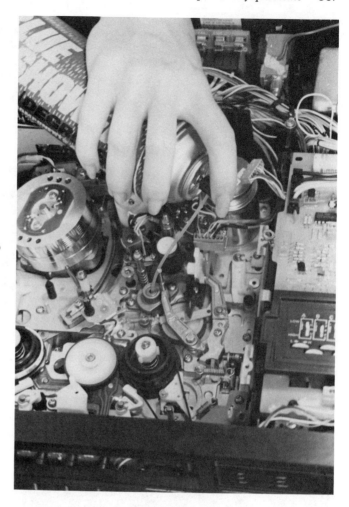

31-9 Spraying degreaser down into the capstan shaft hole.

shaft. Use a dry papertowel to wipe off the capstan shaft. It's important to wipe off any excess oil. If there is any excess oil, it will get on the video tape and clog the video heads.

Squeaking or squealing in play

The VCR is working fine, but has a loud squeak or squeal when in play or record. The bearings in the capstan shaft are dry and need to be lubricated. Follow the same procedure as previously described for lubricating the capstan shaft.

Checking newer direct drive VCRs

Direct drive systems are used in newer models. The flywheel of the capstan shaft is part of the motor. Newer direct drive VCRs have sealed roller bearings and don't need to be lubricated. Get inside the undercarriage. (Refer to chapter 25, "Getting into the undercarriage.")

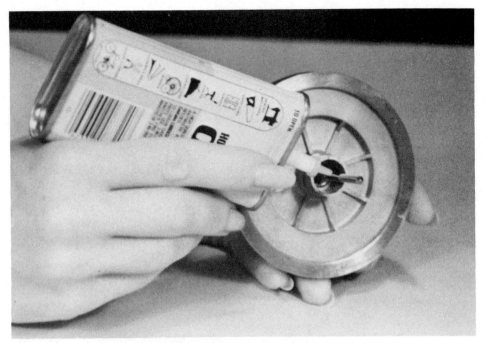

31-10   Lubricating only the lower portion of the shaft.

31-11   Placing grease on the base of the flywheel.

31-12 The flywheel adjustment
screw.

There are three ways to identify these types of units. You'll find a circuit board with wires attached to it protruding out from behind the flywheel, as shown in Fig. 27-6. You also might find a circuit board completely surrounding the flywheel, as shown in Fig. 27-7. Finally, you might find a belt on a pulley attached to the center of the flywheel. If you have a belt on the outside of the flywheel, you have a belt driven VCR. If a direct drive capstan system is malfunctioning, it has a circuit problem and you'll need to take it to a local service center for repairs.

Review

☐ To find out if you have a capstan shaft problem, insert a video cassette and push Play. Check to see if the video tape is moving. If not, check the capstan shaft to see if it's turning.

☐ If you have a direct drive VCR, spin the capstan shaft to see if it spins freely.

☐ If you have an older direct drive VCR and the capstan shaft turns hard or is frozen up, lubricate the oil well at the base of the capstan shaft.

☐ If you have a belt-driven VCR, remove the belt and spin the flywheel to check its freedom of movement.

☐ If the flywheel turns hard or is frozen up, remove the bottom bracket, the capstan shaft, and flywheel for cleaning and lubricating.

☐ Be sure to remove the washer around the capstan shaft before removing the shaft.

☐ Clean the capstan shaft hole with a glass brush.

☐ Be sure to replace the grease on the mounting bracket and the bottom of the flywheel.

☐ Before disassembling the capstan shaft assembly, make sure your VCR is not a direct drive unit.

32

Cassette lamp problems

ALL VCRS MANUFACTURED IN THE FIRST FIVE YEARS DEVELOP A COMMON problem. After inserting a video cassette, everything lights up as normal; however, pushing Play, Fast Forward, Rewind, and Record has no effect. This problem is caused by a bad cassette lamp. A cassette lamp is a small 12-volt light bulb. This bulb burns out periodically and has to be replaced.

Cassette lamp functions

A cassette lamp has two functions. At each end of a video tape there's transparent leader tape. When a VCR is in play, fast forward, or rewind and it reaches the end of the video tape, the light shines through the transparent tape and activates a light sensor. The sensor stops the unit to prevent damage to the video tape. If the bulb burns out, the unit shuts down.

Locating the cassette lamp

To locate the lamp in topload VCRs, turn the power on and push the cassette carriage down into its locked position, without a video cassette. Look thru the plastic window on the cassette lid. You're looking for a small light bulb protruding up behind the bottom plate of the cassette holder, as shown in Fig. 32-1. This bulb is the cassette lamp.

For frontload units, push the Eject button to remove any video cassette. Open the cassette door and look through the door for the cassette lamp at the same location shown for topload VCRs.

Determining the type of cassette lamp

VCRs produced through 1982 are the models using light bulbs for cassette lamps. After 1983, manufacturers started using an infrared bulb, which rarely burns out. Impossible to

32-1 The cassette lamp.

determine whether an infrared bulb is lit because of the type of rays it produces. This device has a bubble on either side and doesn't look like a normal light bulb (Fig. 32-2). If the unit has an infrared bulb, don't attempt to replace it. These types of bulbs extremely rarely burn out. Another problem is causing this symptom. Refer to chapter 14, "Diagnosing VCR problems."

Unplug the VCR and remove the main top cover. Refer to chapter 2, "Getting inside a VCR." There are two different types of bulbs used for cassette lamps, as shown in Fig. 32-3. The one on the right has two blue wires attached. The one on the left has two small pins coming out from the base of the bulb.

Checking the cassette lamp

You need to check the cassette lamp to determine whether you have a burned out bulb or a circuit problem. If the digital clock lights up, the VCR is producing a good picture with sound, and all indicator lights on the front panel light up, then you have a burned out cassette lamp. If any of these functions aren't working properly, then either a fuse has blown or there's a circuit problem. Refer to chapter 12, "Replacing fuses." The majority of the time, however, the bulb has gone bad.

Replacing the cassette lamp

There are two types of cassette lamps. The first type has two blue wires attached to it. The second type has two wire pins protruding from its base. Follow the instructions for replacing the type of lamp that is in your VCR.

32-2 An infared cassette lamp.

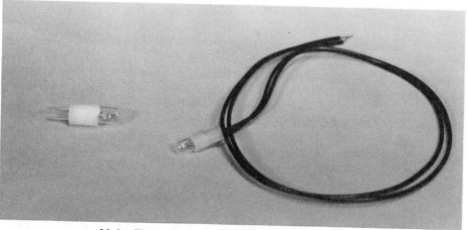

32-3 The two types of bulbs used for cassette lamps.

First type

The bulb with two blue wires attached to it is mounted inside a black plastic cover with a small port hole on each side, as shown in Fig. 32-4. To remove this type of bulb, remove the two Phillips head screws on each side of the black plastic cover. Take hold of the cover and pick it straight up. Insert the tip of a small flathead screwdriver or the tip of a fiberglass brush

32-4 A cassette lamp inside a plastic cover.

through one of the port holes in the cover. Placing the tip of the tool on top of the bulb, pry the bulb further down into its cover (Fig. 32-5). Then, flip the cover over so the wires are facing toward you. Take hold of the wires and pull the bulb out of the cover, as shown in Fig. 32-6.

Follow the path of the two blue wires coming from the bulb. When removing these wires from their path, make note of exactly how they're placed through the transport. The wires go through a wire clamp and through a hole in the chassis along with many other wires. It's hard to follow the wires from this point. Go back to the wire clamp, unhook it, and remove the blue wires. Cut both wires approximately three inches from where they disappear into the chassis, as shown in Fig. 32-7. (Be sure to unplug the unit before you cut the wires.) Take the bulb with the wires, along with the make and model number of your VCR, to a local electronics supply center for replacement.

Take the new bulb and push it up inside through the hole in the bottom of the black plastic cover. Be sure to insert the bulb all the way up, so the light from the bulb shines through both port holes, as shown in Fig. 32-8. There's a flange on the bottom of the bulb to secure the bulb to the plastic mount. Next, place the wires through the notch at the base of the plastic cover (Fig. 32-9). While holding the wires in the notch, insert the mounting screws and remount the plastic cover.

Next, take the two wires coming from the bulb and place them across the transport in the same path as before. Be sure the wires are clear of all moving parts. Bring the new wires up to the old wires. Go past the old wires approximately 1 inch and cut the new wires off at this point. The extra inch of wire gives you enough slack to work with. Remove approximately ½ inch of insulation off all four wires. Solder together the new and old wires. It makes no difference which two wires are soldered together.

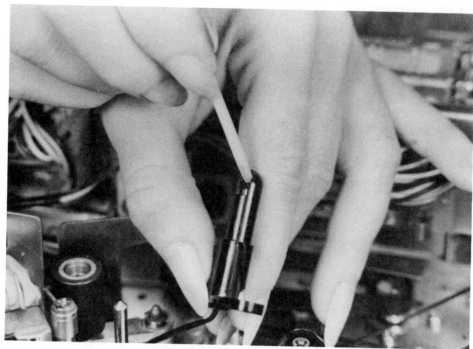

32-5 Prying out the cassette lamp.

32-6 Pulling the bulb out of the cover.

32-7   Cutting the cassette lamp wires.

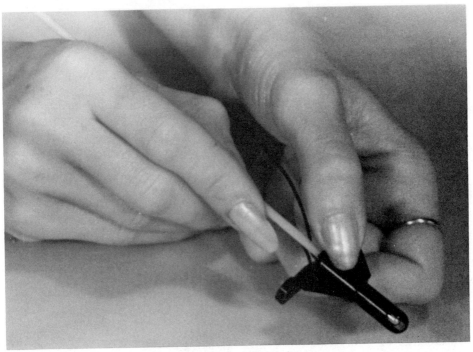

32-8   Inserting the bulb into its cover.

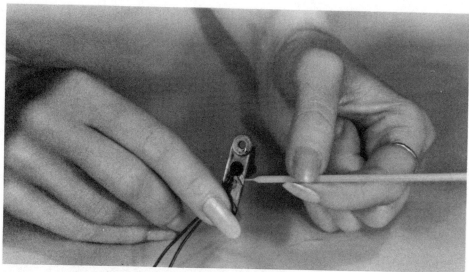

32-9   Placing the wires through the notch at the base of the cover.

Preheat a soldering gun or iron. Then, coat the tip of the iron with solder and flip off the excess solder. Place the solder and iron onto the exposed wire. Melt some solder onto the wire. Remove the solder, leaving the iron on the wire and rubbing it back and forth. This procedure leaves a coating of solder on the wire, allowing the two wires to bond together. Follow this procedure for each wire.

Take hold of one new wire and overlap the exposed end with one of the old wires, as shown in Fig. 32-10. Place some solder onto the tip of the soldering iron. Place the tip of the iron onto the overlap of the two wires, leaving the iron on the connection until the solder is absorbed into the wires. Follow the same procedure for the other two wires.

Place black electrical tape over each soldered connection. Tape each connection separately, so you don't accidentally short two wires together. Replace the excess wire back into the wire clamp, as shown in Fig. 32-11.

Plug the unit in and push Power. The cassette lamp now is working. You also will discover that all the functions have returned to normal. In a few topload models, you will need to push down the cassette carriage to its locked position, then the cassette lamp lights up.

Second type

The most common type of lamp has two wire pins protruding out from the base of the bulb, as shown in Fig. 32-3. To replace this type of bulb, locate the small circuit board under the cassette lamp. This circuit board usually is mounted to the transport by one mounting screw. There also is a small wire clamp attached to this mounting screw (Fig. 32-12). Remove the mounting screw and wire clamp. Lift the board straight up and turn it over. Notice where the two pins from the bulb go through the circuit board and come out on the opposite side. Use a solder wick to remove the old solder. Refer to the section in chapter 24 on "Using a solder wick to remove old solder."

After removing the old solder, turn the circuit board around and remove the bulb. In some models, there is a black lamp mount at the base of the bulb. This mount comes off with

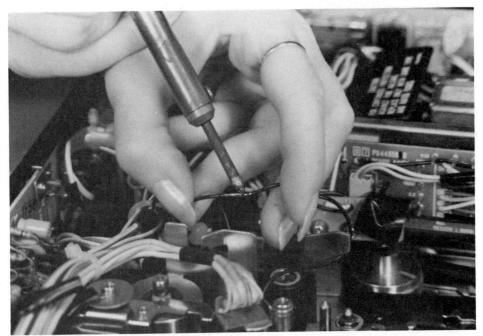

32-10 Overlapping and soldering the two wires.

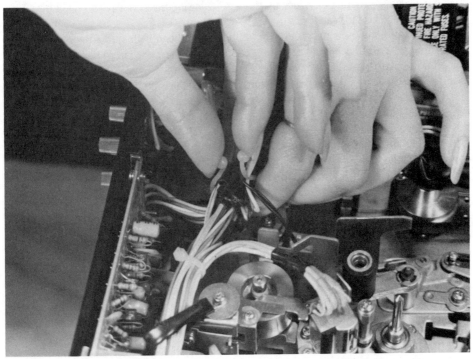

32-11 Placing the excess wires into the wire clamp.

32-12   The wire clamp and the cassette lamp mounting screw.

the bulb. To remove the mount from the bulb, take hold of the mount and pull the bulb out, as shown in Fig. 32-13. In other models, there is no mount to remove.

When replacing the lamp and mount to the circuit board, insert the pins of the bulb through the mount. Make sure the pins are protruding through the small holes in the bottom of the mount. Look at the bottom and you'll find two plastic guide pins. When replacing the bulb, make sure these guide pins and the pins from the bulb are aligned with the holes in the circuit board, as shown in Fig. 32-14. To solder the pins to the circuit board, refer to the section in chapter 24 on "Soldering pins to a circuit board."

Before remounting the circuit board, look at the transport for a mounting pin (Fig. 32-15). Line up the hole on the circuit board with the pin on the transport. The other hole on the board is for the mounting screw. Insert the mounting screw through the wire clamp and the circuit board. Tighten the screw down.

Plug in the unit and push Power. The cassette lamp should light up, returning all functions to normal.

Review

☐ If the unit lights up but won't go into play, rewind, or fast forward, check to see if the cassette lamp is lit.

☐ If the bulb is not lit, make sure the cassette lamp isn't an infrared device.

☐ Unplug the unit.

☐ Remember there are two types of cassette lamps:
  • One has wires attached that have to be cut to remove the bulb.
  • The other has wire pins and has to be unsoldered to be removed.

☐ When replacing the bulb, be sure not to solder the neighboring connections together.

☐ If the light in some models does not light up after the bulb has been replaced, push the cassette lid down to its locked position.

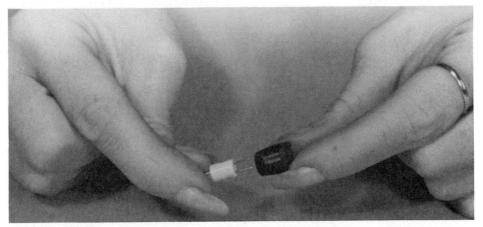

32-13   The cassette lamp and the black mount.

32-14   Aligning the lamp pins to the holes on the circuit board.

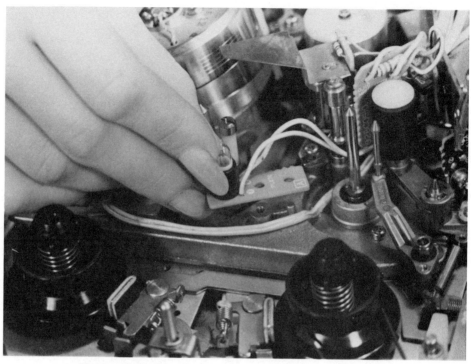

32-15   Aligning the circuit board hole to the pin on the transport.

33

Older VCRs jamming due to power outages

THE FIRST VCRS PRODUCED WERE BELT DRIVEN. THESE VCRS ARE COVERED in this chapter. Belt-driven models have a major flaw. If there's a power outage or the unit is accidentally unplugged, the unit is stuck in play when the power is restored but the unit doesn't produce a picture. Pushing the mode buttons has no effect, and the video cassette won't eject.

The cause

When the power is lost, the unit automatically unloads. When the power is restored, the unit is still unloading. About halfway through this mode, a lot of tension is put on the drive belt. The flywheel creates enough initial force to keep it unloading. When the power is lost, the flywheel loses its initial force, causing the unit to jam. If your unit jams, turn the unit off.

Locating the problem

Remove the main top cover. Refer to chapter 2, "Getting inside a VCR." Turn the unit on and look on top of the transport for the drive wheel that's spinning. The belt around the wheel (Fig. 33–1), however, isn't moving. This belt is the main drive belt, which drives the rest of the unit. The pulley that's spinning is directly attached to the capstan motor and drives the main drive belt.

In most models that develop this type of problem, the main drive belt is on the top right side of the transport. In other models, the main drive belt is in the undercarriage. Refer to chapter 25, "Getting into the undercarriage."

Correcting the problem

While the capstan motor drive wheel is spinning, take your finger and place it on the opposite drive pulley, as shown in Fig. 33-2. Give the pulley a quick spin clockwise until it starts turning on its own. The unit completes the unloading process and stops. You'll find all the functions are back to normal.

To keep this problem from reoccurring, clean or replace this drive belt. Refer to the first four sections in chapter 26 for instructions.

33-1 The drive wheel.

33-2 The drive wheel pulley.

34

High-pitched sound coming from a VCR

WHILE VIEWING A MOVIE, YOU SUDDENLY HEAR A HIGH-PITCHED SOUND OR squealing coming from the VCR. The sound isn't coming from the speakers in the monitor. When the sound first occurs, it periodically comes and goes, but only in play or record. Later, the sound becomes fairly constant. When you stop the unit, the sound continues for another one to three minutes.

Locations where a high-pitched sound will occur

There are three different locations at which you might hear a high-pitched sound. Each location is covered in this section and has its own set of instructions. To find the location from which the sound is coming in your VCR, read each of the following sections.

First location

Remove the main top cover. Refer to chapter 2, "Getting inside the VCR." Insert a video cassette and push Play. As soon as you hear the high-pitched sound, look at the top of the video drum. You'll find a grounding leaf spring, as shown in Fig. 34-1. The leaf spring is producing this sound. While the unit is producing this sound, push lightly on the spring. As soon as you touch the spring, the high-pitched sound or squeal should stop. Although the sound has stopped, this doesn't correct the problem. The sound is caused by a lack of lubrication.

Second location

In some models, the grounding leaf spring is in the undercarriage. Place the unit on its side. Refer to chapter 25, "Getting into the undercarriage." Place your finger on top of the video drum and spin it. Look in the undercarriage for a small flywheel that's spinning. At the base of the flywheel is a leaf spring, as shown in Fig. 34-2.

34-1   First location for a grounding leaf spring.

34-2   Second location for a grounding leaf spring.

34-3   The two mounting screws for the rotary cap.

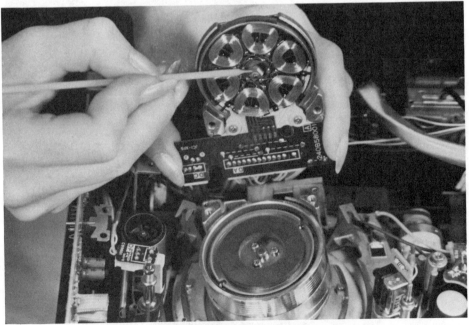

34-4   Third location for a grounding leaf spring.

34-5 A pin on the bottom of the rotary cap.

Third location

In other models, the grounding leaf spring isn't accessible. These types of units have a rotary cap on the top of the video drum. The cap must be removed to reach the spring. There are two mounting screws located on top of the rotary cap, as shown in Fig. 34-3. Remove the screws and pick the rotary cap straight up and off. There are magnets inside holding it down, so it'll be hard to pick up. After removing the cap, you'll see the leaf spring (Fig. 34-4).

When remounting the rotary cap, you'll notice two small pins protruding from the bottom. On top of the mounting bracket behind the video drum are two pinholes beside each screw hole, as shown in Fig. 34-5. When replacing the rotary cap, line up the two pins with the pinholes. The magnets cause the cap to clamp against the top of the video drum. Insert and tighten down the two mounting screws, automatically aligning the rotary cap to the video drum.

Removing the high-pitched sound

Stop the unit and allow the video drum to stop spinning. Saturate a chamois stick with cleaning alcohol. While slightly lifting up on the spring, slide the chamois stick between the top of the spring and the top of the video drum. Keeping a slight amount of pressure on the top of the spring, rub the chamois stick back and forth, as shown in Fig. 34-6. Reposition the cham-

34-6   Cleaning the top of the video drum's shaft and the bottom of the spring.

34-7   Lubricating the shaft and its leaf spring.

ois stick to remove all the black residue. You're cleaning the top of the video drum shaft and the bottom of the spring.

The grounding leaf spring removes static electricity from the video drum. If the spring is improperly cleaned, white dashes appear in the picture. Place some phono lube on the tip of a small flathead screwdriver and lift up the leaf spring. Place the phono lube on top of the video drum shaft and the bottom of the spring, as shown in Fig. 34-7.

35

Repairing broken
or nonfunctional
operating buttons

THE VCR WAS ACCIDENTALLY BUMPED OR DROPPED ON THE FRONT COVER. A function button has broken off, is loose, or no longer functions. You would assume each button is a separate piece, but most buttons are manufactured right into the front cover. A VCR repair shop would probably replace the entire front panel cover, which is expensive and unnecessary.

First steps in removing a front panel

Remove the main top cover. Refer chapter 2, "Getting inside the VCR." The top of the front cover is mounted in one of three ways.

- Two to four Phillips head mounting screws across the lip on top of the front cover.
- Three to four locking latches across the top inside of the front cover with a screw in each one, as shown in Fig. 35-1. Some models only have two screws.
- Three to four locking latches across the top inside of the front cover with no screws, as shown in Fig. 35-2.

Remove any screws on top of the front cover. Open up any panel doors located on the front of the unit. Some models have a hidden screw inside a door. If so, remove it. Lay the unit on its side and remove the bottom cover plate. Refer to the section in chapter 25 on "Removing the bottom cover plate." In all models, these are the first steps to follow.

Removing front covers

There are two different types of front covers on VCRs: covers with locking latches and covers with mounting screws. The following sections provide instructions for removing both types of covers. Check your VCR to determine which type of cover your unit has.

35-1   A locking latch with its screw inserted.

35-2   A locking latch with no screw.

Covers with locking latches

Look for locking latches on the top, ends, and bottom of the inside of the front cover. These latches have a round plastic pin protruding through a hole in the latch (Fig. 35-3). In other models, the latches have a rectangular shape pin protruding through a hole in the latch (Fig. 35-2).

To release the latches, take hold of a latch and pull it out until it clears the pin protruding through the hole. Then, pull out on that section of the front cover to keep the latch from recatching (Fig. 35-4). Proceed with each latch in the same manner until the front cover comes off.

If you only find latches on the top of the front cover, the unit has three to four hidden latches. To remove this type of front cover, release the top latches and pull the top of the cover forward, releasing the bottom latches. Figure 35-5 shows a hidden latch. Each hidden latch consists of a slot on the chassis and a hook on the front cover. The hook slips into a slot to hold on the bottom cover. In other models, the hook is on the chassis and the slot is on the bottom of the front cover. The principle, however, remains the same.

Covers with mounting screws

There are no latches in older models. The front cover is mounted by two to six Phillips head screws. There are two locations for these mounting screws: across the top and bottom of the front cover or on each end of the front cover. Just remove the mounting screws and pull the front cover straight off.

35-3 A plastic pin protruding through the latch.

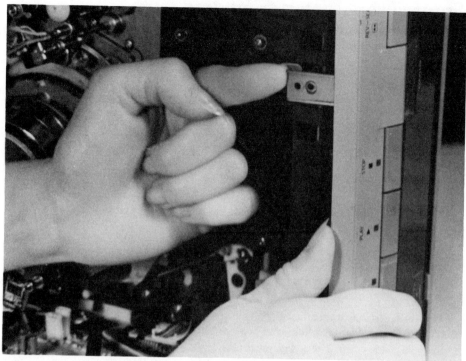

35-4   Pulling out on the cover to keep the latch from recatching.

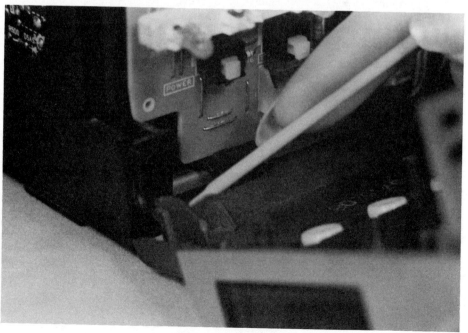

35-5   A hidden latch.

Front covers with a grounding wire

In a few models, after removing the front cover, you'll find a black wire attached to it. This black wire grounds the middle of the front cover to the chassis. The wire is always attached to the front cover by a small plug. Simply unplug the wire from the cover. Most models don't have a grounding wire. Just be aware of the possibility.

Remounting front covers

First, align any slide controls or switches with knobs mounted to the front cover. If the unit only has mounting screws, push the cover straight on and insert the screws. If the cover has hidden latches, insert the hook into the slot at the bottom of the front cover and push the top cover shut like a door. Continue pushing until each latch pops into place. Then, insert and tighten the mounting screws. If the unit has latches on the top, bottom, and sides, push the cover straight on until each latch pops back into place. Then, insert and tighten any mounting screws.

Locating and repairing missing buttons

The missing button usually is floating around inside of the front cover. Most buttons have a lip around their base to keep the button from falling out.

Buttons break off at their hinge. The hinge usually is made of two plastic fingers at the base of the button. Place the front cover on its face. Reposition the button back through its proper hole on the cover. Then, align the broken fingers on the hinge.

There are two methods used for repairing buttons. I highly recommend using the clear silicone glue method. Apply clear silicone glue to the hinge. Completely cover both fingers of the hinge with the silicone, as shown in Fig. 35-6. The silicone glues the hinge but remains

35-6 A button repaired with silicon glue.

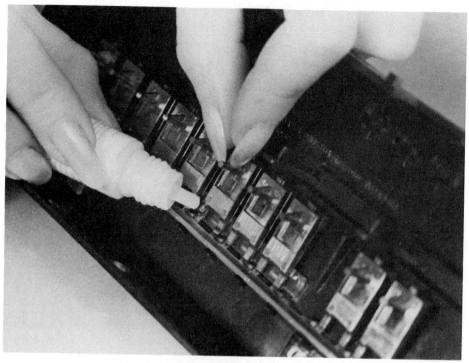

35-7   Repairing the broken fingers on a hinge.

flexible. If the hinge is partially or completely missing, use the silicone glue to form a new one. Use the silicone liberally. The bonding is what's important, not the appearance.

Another method for repair is to place a small drop of super glue onto the break of the hinge, as shown in Fig. 35-7. Hold the button in place until the glue is set. Use a piece of fiber tape or black electrical tape to hold it into position permanently, as shown in Fig. 35-8. The hinges on the button must remain flexible.

Sometimes a button is held on with a shaft instead of a hinge. This shaft breaks right at the bass of the button. The button falls outside of the VCR instead of the inside leaving the shaft protruding through the button hole (Fig. 35-9). Drop a bead of super glue onto the end of the shaft and place the button up against the shaft, holding them together until the glue is set.

Repairing twisted buttons

If a function button is crooked or out of alignment, one of the fingers is broken on the hinge. Twist the button back into position and place a drop of super or silicone glue onto the broken finger. Hold it in place until it's set.

Why mode buttons become nonfunctional

A button is nonfunctional if it's intact but doesn't function. Three different problems can cause this to happen. First, the shaft behind the button has broken off, as shown in Fig. 35-10. If

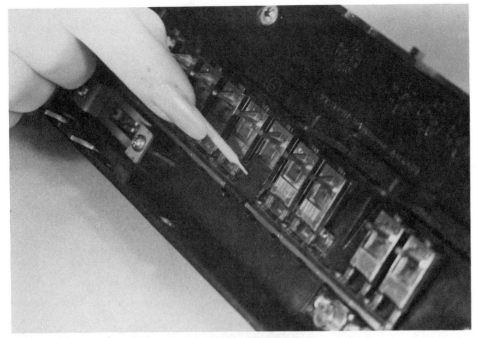

35-8 Taping a button to hold it into position.

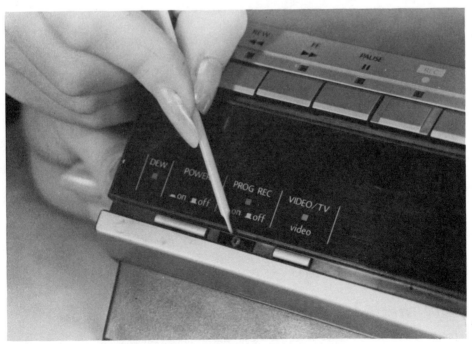

35-9 A shaft protruding through the button hole.

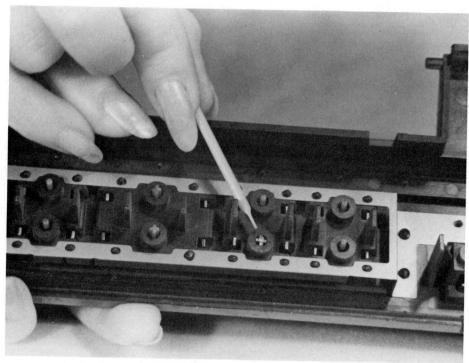

35-10   A broken off button shaft.

you can find the broken piece of shaft, simply super glue it back into place. If you can't find the piece of shaft, follow this procedure: Remove the empty ink well from a clear pin and cut a section off that is the same length as the other button shafts. Then, super glue it to the broken shaft.

The second cause is a broken microswitch. Each button mounted to the front cover has a shaft protruding out the back of the button (Fig. 35-11). When pushing on a button, the shaft on the back of the button pushes against a microswitch. The microswitches are located directly behind each button on a circuit board (Fig. 35-12). These microswitches fall apart if impacted too hard. Although the microswitch has been broken, the button on the front cover might be intact.

Incorporated into the center of the microswitch is a little plastic button. On impact, the button pops out, leaving the switch nonfunctional (Fig. 35-13). In this case, the microswitch has to be replaced.

The third cause is a cracked circuit board. Proceed to chapter 36, "Repairing a VCR that has been dropped."

Replacing microswitches

Unplug the unit and remove the circuit board the microswitch is mounted to. Proceed to the section in chapter 36 on "Boards behind the front cover" under the subhead "Removing circuit boards." After removing the circuit board, turn it over. With the back side facing you, locate the broken microswitch and the pins protruding through the back of the board. To aid

35-11   The shaft protruding out the back of the button.

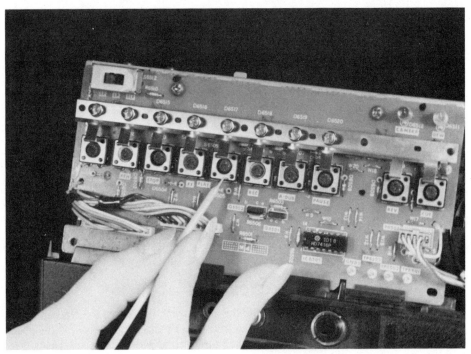

35-12   A microswitch on a circuit board.

35-13 The missing button on a microswitch.

in identification, a black or white square outline is around each set of microswitch pins, as shown in Fig. 35–14. You'll find four or five pins within each square.

Where the pins come through the board, place solder wick over one of the soldered connections. Refer to the section in chapter 24 on "Using a solder wick to remove old solder." After removing the solder, turn the board over and pull out the microswitch. Take the old microswitch, along with the make and model number of your VCR, to a local electronics supply center for a replacement.

To replace the microswitch, place the pins on the microswitch through the holes on the circuit board. The pins on the microswitch are offset, so a switch can't be put in backwards. Turn the board over. Refer to the section in chapter 24 on "Soldering pins to a circuit board." Be careful not to solder any pins together. If you do, use solder wick to remove any excess solder.

Replacing slide switches

A slide switch (Fig. 35-15) also stops functioning upon impact. Slide switches are used for setting the tape speed, stereo hi-fi, audio inputs, and so on. On impact, the switches fall apart. Remove the circuit board attached to the switch. To replace these switches, use the same method as previously explained for microswitches. The only difference is the number of pins. These switches have anywhere from 4 to 12 pins per switch. The pins have a rectangular outline around them, as shown in the lower left corner of Fig. 35-14. Inside this outline is where the pins for the slide switch are located.

35-14 The special white marking, indicating each microswitch.

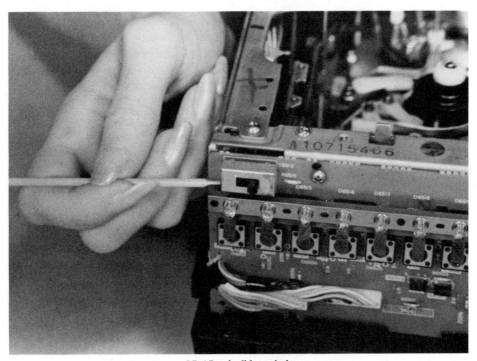

35-15 A slide switch.

Review

- ☐ Remove the main top cover.
- ☐ Remove the bottom cover plate.
- ☐ Remove any screws on the inside lip of the top of the front cover.
- ☐ Check all panel doors for a hidden screw.
- ☐ Release all latches around the front cover.
- ☐ If there are hidden latches, pull the top of the front cover toward you, and the bottom will release.
- ☐ Properly position the broken button.
- ☐ Use silicone glue to repair any parts on a broken hinge.
- ☐ The hinge must remain flexible.
- ☐ If a button is nonfunctional, check for a broken microswitch.
- ☐ Remove all switches by unsoldering their pins.
- ☐ Be sure not to solder two pins together when replacing a switch.

36

Repairing a VCR that has been dropped

WHEN YOU DROP THE VCR, ONE OF THREE THINGS CAN HAPPEN. ONE OF THE outside plastic covers cracks or breaks. Some of the mode buttons stop functioning. In playback, the picture is lost or washed out, or there is no picture or sound. The unit also can go completely dead.

Another problem a VCR can encounter is improper handling in shipping. The unit was improperly packed in its box, and by the time it reaches its destination, it's nonfunctional.

Why a VCR stops working

The majority of the time when a VCR has been dropped, a circuit board inside cracks or breaks. A circuit board is the most brittle part inside a VCR. The printed circuits break apart, causing the unit to malfunction. There are anywhere from one to nine circuit boards in a VCR. Most of the circuit boards are located right inside the outside covers, surrounding the main chassis. Usually, it's the boards surrounding the main chassis that crack or break on impact.

Locating cracks

Locate the point of impact on the outside cover. Unplug the unit and remove the outside cover. Refer to chapter 2, "Getting inside the VCR," chapter 25, "Getting into the undercarriage," and the first five sections in chapter 35. Locate the circuit board closest to the point of impact.

Between the circuit board and the outside cover are small bumpers made of plastic or rubber that help support the circuit board. They're mounted to the interior of the cover or on a circuit board. On impact, the outside cover pushes the bumper against the circuit board, causing the board to break.

Another location to check is under or around any mounting screws. The mounting screws hold the circuit board in place. If the unit fell on its face, check around all the function buttons and switches. If the unit is completely dead, check the board the power transformer is connected to.

Cracks on a bottom board

After removing the bottom cover plate, check the inside of the cover for bumpers. There are two types of bumpers. One is made of rubber and is mounted to the inside of the cover, as shown in Fig. 36-1. The second is made of white plastic and is mounted to the circuit board, as shown in Fig. 36-2. In other units, the white bumpers are mounted to the cover itself. If the bumpers are mounted to a cover, check the location between the bumper and the point it makes contact with the circuit board. If the bumpers are mounted to a circuit board, check around each bumper for cracks. Also, check around each mounting screw.

Cracks behind the front cover

Remove the front cover and look inside the unit for plastic bumpers, as shown in Fig. 36-3. (The bumpers might be mounted to the circuit board instead.) Look inside the cover at the location of each bumper. Compare the contact point of each bumper to the circuit board. If a VCR is dropped on its face, the circuit board will be broken at that point. If the bumpers are mounted to the circuit board, check around each bumper. Check around each mode button, slide switch, and the mounting screws.

36-1 A rubber bumper.

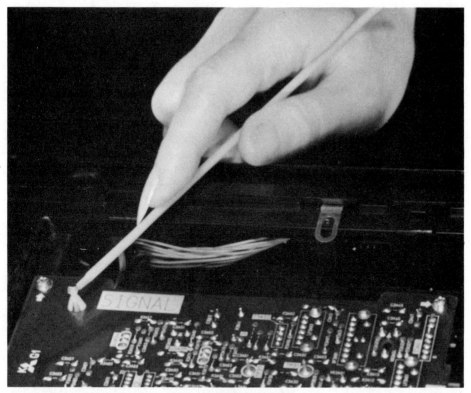

36-2   A white plastic bumper.

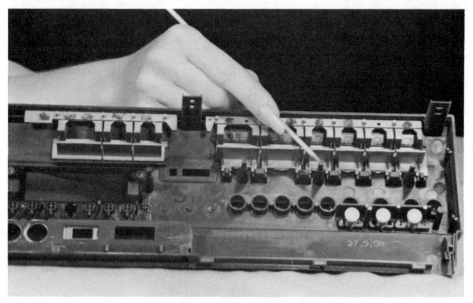

36-3   A plastic bumper.

36-4   A circuit board mounting groove.

Cracks on the side or top of VCR

Remove the main top cover to locate cracks on boards on either side of the VCR. Some circuit boards are mounted by grooves. The board slides down into a groove, with one on each end of the board. The grooves hold the circuit board in place. (See Fig. 36-4. I've pulled the board up out of the groove to give you a better view.) The grooves differ in length. When a unit falls on its side, the cover pushes against the circuit board and causes the board to bend. The grooves don't bent; consequently, the board breaks right at the groove. If the board is mounted to the chassis with mounting screws, the cracks appear around the screws.

To locate cracks on boards on top of the VCR, look around all the mounting screws. If the board has hinges, look around each hinge. Figure 25-9 shows the top hinges.

Cracks on a board mounted to a power transformer

If the unit goes completely dead, see if the power transformer is directly connected to a circuit board. Then, check the board for cracks. Figure 36-5 shows where the crack can appear on the circuit board. Directly on the other side of the board is the power transformer. On impact, the weight of the power transformer breaks the board. Next, check around all the mounting screws for cracks. All other boards are located in the middle of the unit and usually won't break or crack on impact.

Removing circuit boards

The steps required to remove a circuit board vary depending on the location of the board. The next five sections give instructions for the various locations. Follow the instructions in the section that pertains to your situation.

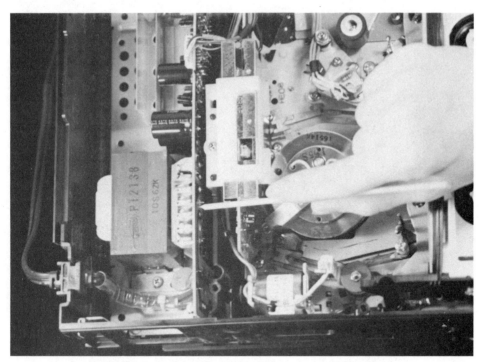

36-5 A circuit board connected to a power transformer.

Boards under the bottom cover

To remove any circuit boards under the bottom of the VCR, refer to the sections in chapter 25 on "Removing the bottom plate" and "Bottom circuit boards."

Boards behind the front cover

Remove the front cover. There are between one and three circuit boards mounted to the front of a chassis. These boards might be separate or connected together.

In some models, the latches are located on the bottom of the board, as shown in Fig. 36-6. There are one to six latches, which vary in size. In some models, these latches protrude through small square holes in the circuit board. Starting at either end, pull the latch away from the board, simultaneously pulling that section of board away from the chassis. Proceed by doing each latch separately in the same manner. After all of the latches have been released, pull the bottom of the board out toward you. Pull the board down to clear each lip across the top of the board, as shown in Fig. 36-7. Then, place the board flat on the table for repairs.

In other models, the releasing latches are on top of the board, and the lips on the bottom. Simply release the latches on top and pull the top of the circuit board towards you, as previously explained. Lift up on the board to clear the lips on the bottom. Most boards have extended leads attached to the circuit board. You don't need to unplug the circuit board; just stretch the leads out and lay the circuit board onto the table for repair.

Another model has screws across the top or bottom of the circuit board, as shown in

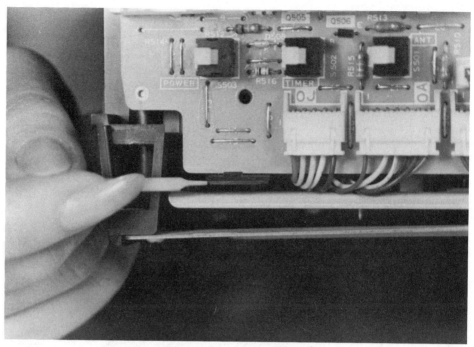

36-6   A latch at the bottom of a circuit board.

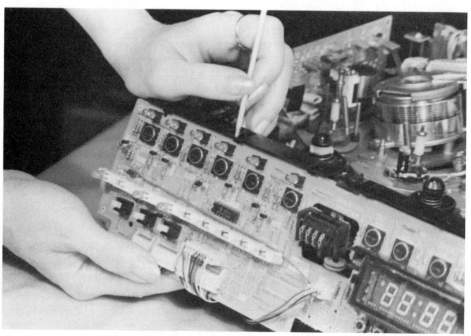

36-7   Pulling down on the circuit board to clear the lips.

36-8 The circuit board mounting screws.

Fig. 36-8. Remove the screws and slide the board in the direction of the screws, clearing the groove across the opposite end of the board. Lay the board flat on its face for repairs.

A different model has mounting screws completely around the circuit board. Remove the screws and pull the board straight off. Lay the board flat on its face for repairs.

In another model, the circuit boards on the front of the chassis are connected to the circuit board on the bottom of the chassis. Refer to the section in chapter 25 on "Sixth way" under the subhead "Ways to open a circuit board."

After pulling the circuit board out, look at the chassis. You'll find plastic bumpers supporting the back of the circuit board (Fig. 36-9). These bumpers also can cause the board to crack. Check the circuit board at these locations. You'll find two sets of bumpers one in front of and one behind each circuit board. All other boards have bumpers on one side or have none at all.

Boards on either side

There are three ways to remove boards on the sides of your VCR. Pull the board straight up and out of the groove on the bottom of each side. (Refer back to Fig. 36-4). If the boards are mounted to the chassis by mounting screws, remove the screws and pull off the board. In other models, a white latch with a pin is protruding through a hole in the board near the top, with one latch on each side of the circuit board. Pull the latch back while pulling the board towards you, as shown in Fig. 36-10. Repeat this procedure for the other end. The board opens like a door (Fig. 36-11).

36-9   Plastic bumpers behind the circuit board.

36-10   Releasing a white latch.

36-11   Opening a circuit board like a door.

A board connected to a power transformer

Before removing a board connected to a power transformer (Fig. 36-5), you must unplug the VCR. Electrical shock is possible. There are two ways to remove this type of board. In both cases, the power transformer comes out with the circuit board. The first type has two Phillips head mounting screws at the base of the transformer that are screwed into the plastic part of the chassis. Remove these screws. Pull the power transformer and circuit board straight out.

The second type of board has the power transformer and the circuit board mounted to a metal shield under the transformer. You'll need to remove the mounting screws from the shield. You'll find three to four mounting screws, usually in each corner of the shield. Sometimes there is a screw in the middle. Two screws are at the base of the transformer. These two screws hold the transformer to the metal shield. (I've pointed out the mounting screws with arrows in Fig. 36–12.) Remove the screws and pick up the entire power supply, as shown in Fig. 36–13. Some screws have black grounding wires attached.

You'll find one to three white plugs attached to this board. To make the repair easier, detach the plugs. Each plug has a different number of pins. When replacing the board, you'll have no difficulty matching which plug goes where. The ac cord is attached to this board and comes up with it. In some models, one screw is mounted to the bottom cover plate under the power supply. Take out the screw to remove the power supply. Another model has four screws at the base of the transformer with black grounding wires attached. Remove all screws to remove the metal shield.

36-12 Arrows pointing at the mounting screws.

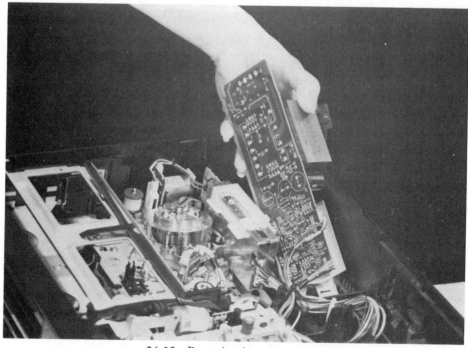

36-13 Removing the power supply.

Boards underneath the top cover

Refer to the section in chapter 3 on "Removing circuit boards blocking the tape path." For any board on the right side of the unit, the mounting screws are red or gold in color, are held down with black plastic clips around the edges, or both. In a few models in addition to the mounting screws, there is a double white clip in the front right corner inserted through a square hole on the board. Squeeze the white clip together and pull up the board. In some models, part of the back comes up with the board. If the components are facing down, the board is not hinged. If the components are facing up, the board always has hinges. In one or two models, the leads coming from the video head hold the circuit board down. Look on the component side of the board for a metal box with a snap-on lid. Pop the lid off and disconnect the plug inside.

Making the repairs

The following sections will show you how to repair cracks to any of the circuit boards in your VCR. Read each of the sections before you begin, then carefully follow the instructions.

Repairing cracks on circuit boards

On the component side of a circuit board, you usually will find a fine white line or branches of lines, as shown in Fig. 36-14. These lines can be difficult to detect. If the board is broken, the crack will be obvious and the board probably will be crooked, as shown in Fig. 36-15. Realign the break in the board so it's flush, leaving just a crack in the board. On the component side only, run a bead of super glue along the full length of the crack to prevent it from spreading (Fig. 36-16). Hold the board together until the glue sets.

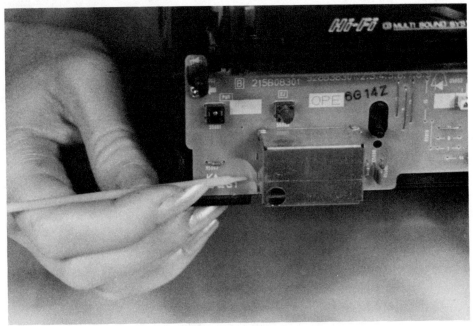

36-14 A fine white crack on the component side of a circuit board.

36-15   A broken circuit board.

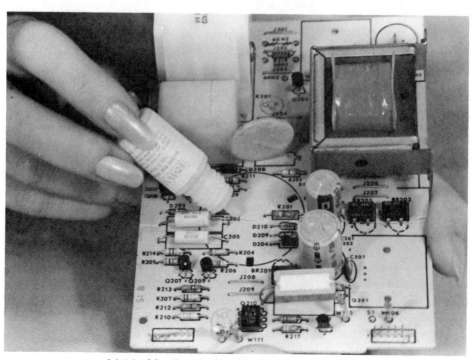

36-16   Mending a board with glue on the circuit side.

36-17   A cracked circuit board.

Figure 36-17 shows how a crack appears on the opposite side, the circuitry side, of the board. Look along the crack. Notice the several points where the circuitry tracks have broken and no longer make contact. This interruption has caused certain functions to stop working. To repair these broken tracks, solder a wire across the broken track.

To solder a wire to a circuit board, remove about ¼ inch of insulation off a strand of 24-gauge wire, then coat the end with solder. Place the coated wire on top of a connection. Then, put the hot solder iron on top of the wire and push a small amount of solder onto the connection. Remove the solder, leaving the solder iron on the connection for a second or two. It's important not to let the solder run into neighboring connections. If you do, the circuit will short out. If you get solder on other connections, use the solder wick to remove the solder. Refer to the section in chapter 24 on "Using a solder wick to remove old solder." Follow the same procedure for soldering wires to circuit boards.

Repairing a broken track

Sometimes the cracks branch out, as shown in Fig. 36-18. The branches are hairline cracks and are difficult to see. Be sure to locate the entire crack and all the cracks. Starting at one end, follow the crack until you come across the first broken circuit track. Tracks are part of the circuitry and connect the components together. The tracks are bonded to the bottom of the board and are like thin tin foil. They are covered with a green or brown colored plastic coating, which is an insulation. As soon as you've located the first broken track, follow the track until you come to the first solder connection. A solder connection is where a component protrudes through the circuit board and is soldered to the track. Take the solder-coated wire

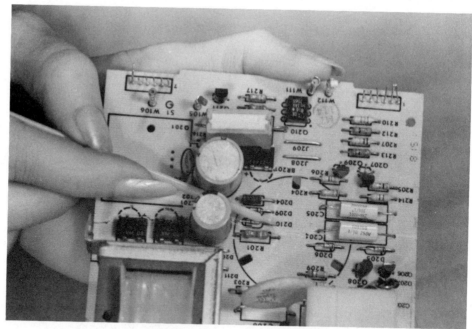

36-18   The crack branching out.

and solder it to this first soldering connection. Figure 36-19 shows the end of a track at the break and at the solder connection.

After soldering the wire to its first soldering connection, retrace the track back to the break and continue on the other side of the break until you come to a soldering connection. Take the wire you soldered to the first connection and stretch it tightly across to the second connection, adding approximately ¼ inch of wire. Cut and remove the insulation off the wire. Then, solder that end of the wire to that soldering connection. Go back and retrace the track to make sure the wire is soldered to the same track on both sides of the crack (Fig. 36-20). The tracks are close together, so it's easy to get them confused.

There can be a long distance between soldering points. Look at Fig. 36-21. Notice how far away the white wire is from the track and the odd location of the soldering connections found on each side of the crack.

Repairing double-printed circuit boards

A few models have double-printed circuit boards. The circuitry side of the board looks the same as previously shown, but the component side also has tracks around the components. On this type of board, you must repair the tracks on both sides of the board. It's important to stay on the right track. Crossing a track can cause more damage to the unit. Recheck all the tracks before plugging in and turning on the unit.

Repairing broken covers

The easiest way to repair a broken cover is with super glue. Take a tube of super glue and run a bead of glue down the crack or the broken piece on the inside of the cover only. Push the two pieces together and hold them together until the glue set.

36-19   Pointing at the track and its closest soldering connection.

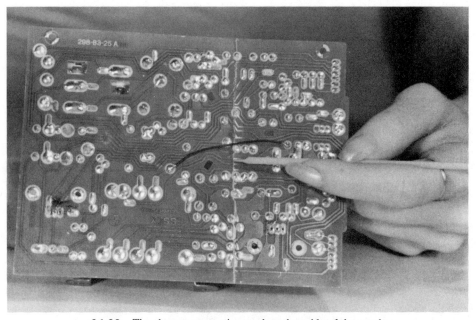

36-20   The closest connection on the other side of the crack.

36-21 A longer distance to the closest soldering connection.

Review

☐ Look for the impact point.

☐ Open the unit and look for the circuit board closest to the impact point.

☐ Remove the circuit board and look for a crack, which can be found around the mounting screws or where a bumper sits up against a board.

☐ If the board is crooked, then straighten the board out, making it flush.

☐ No matter if you have a crack or a break, glue it on the component side only.

☐ Trace the broken tracks carefully and solder a wire to each end of the broken track.

☐ Do not cross the tracks.

☐ Be careful not to solder two neighboring connections together and cause a short.

☐ Remember to retrace all connections first before plugging the unit in.

☐ If the unit goes completely dead, check around the power transformer for a cracked circuit board.

37

Repairing
remote controls

THREE COMMON PROBLEMS OCCUR IN REMOTE CONTROLS. ONE IS THAT THE batteries get weak or go bad. Also, if it has been dropped, the circuit board inside can break. Finally, if a beverage has been spilled on it, the buttons will stick down and some or all of the functions stop working. If replacing the batteries doesn't correct the problem, take the remote apart for repairs.

Locating the mounting screws

Remove the battery cover door on the back of the remote and take out the batteries. There are four locations where the mounting screws might be. All mounting screws are on the back of the remote. The first type has one small Phillips head mounting screw in the middle, above the battery compartment (Fig. 37-1). The second type has two Phillips head screws inside the battery compartment (Fig. 37-2). The third type has the mounting screws hidden under rubber foot pads. In most cases, the two pads opposite the battery compartment contain the mounting screws. Take a small flathead screwdriver, and pop out the rubber foot pad, as shown in Fig. 37-3. The last type has four Phillips head screws, with one in each corner, as shown in Fig. 37-4. Remove all the mounting screws.

Removing the back cover

Look at the side of the remote where the top and bottom covers come together. If you find a small indentation on each side of the remote (Fig. 37-5), follow this next procedure. Pull the back of the remote up approximately ¼ inch. Take a small flathead screwdriver and push into the indentation to release the latch. The back cover will pop up approximately ⅛ inch, as shown in Fig. 37-6. Turn the remote around and release the other latch in the same manner. Grab the back of the remote and pull the back cover up to release the front latches releases, as shown in Fig. 37-7.

37-1   The back cover mounting screw.

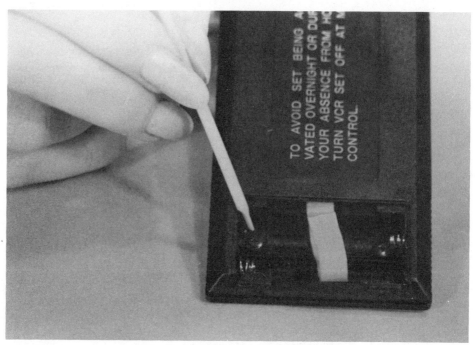

37-2   Two Phillips head mountings screws.

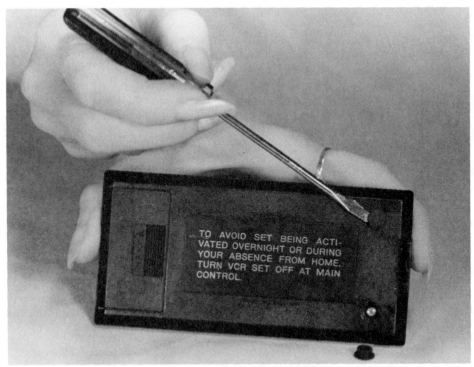

37-3   Lifting the pad to locate the mounting screw.

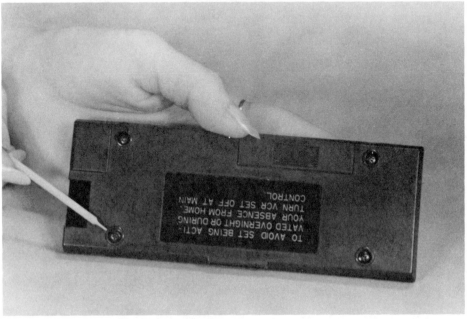

37-4   Mounting screws in each corner.

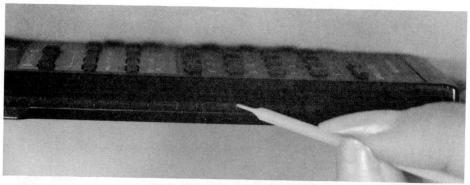

37-5 The indentation on the side.

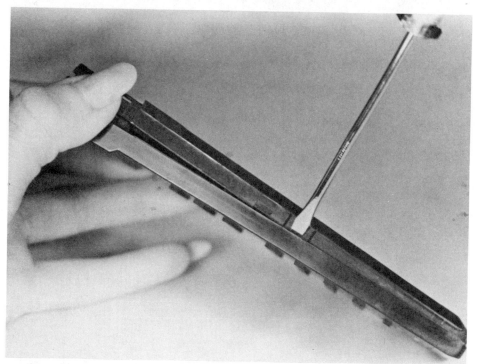

37-6 Releasing the latch.

If there are no indentations on either side of the remote, then follow this procedure. Check for two mounting screws inside the battery compartment. Remove these mounting screws, then slide the bottom cover straight back approximately ¼ inch to release all the latches (Fig. 37-8). The cover then lifts right off.

Another type of remote has no mounting screws or indentations. Take a small flathead screwdriver and shove the tip of it between the top, and bottom cover. Twist the screwdriver and that side will pop up, as shown in Fig. 37-9. Follow the same procedure for the opposite side. Now, you can lift off the back cover.

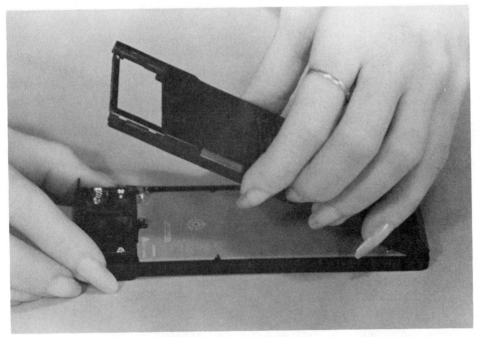

37-7   Releasing the front latches.

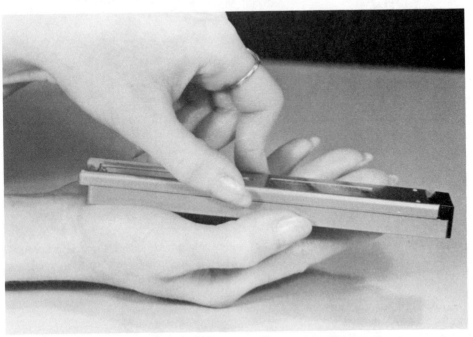

37-8   Pulling straight back to release all the latches.

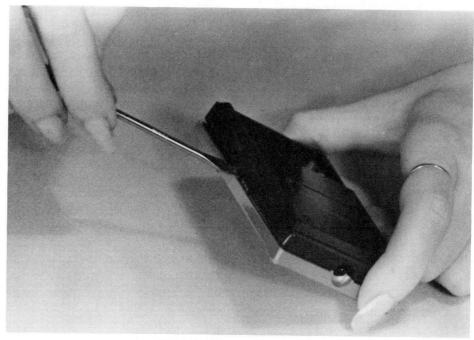

37-9 Popping off the back cover.

The last type has a mounting screw located in each corner of the remote control. Simply remove the screws and lift off the back.

What is an LED?

LED stands for light- emitting diode. A diode is a device that allows electricity to flow through it in only one direction. All remote controls have an infrared device. These devices look like small light bulbs and are called LEDs. These LEDs send out a pulsating infrared light, which signals the VCR what function to perform. Some units have three LEDs, but most have one or two. They face toward the front from behind a red plastic window, which is usually connected to the back cover. A few remotes have no plastic window and the LED protrudes directly through the front (Fig. 37-9).

Removing the window

If the window remained with the front portion of the remote control when you opened the remote, take a small flathead screwdriver, place it at the base of the window, and slide the window up and off (Fig. 37-10).

Battery compartment

In older remote controls, the battery compartment comes off with the back cover. In newer models, the battery compartment is part of the front cover. In either case, you'll find a red

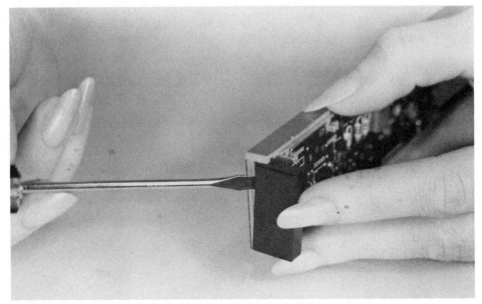

37-10   Removing the window.

and black wire attached to the circuit board, as shown in Fig. 37-11. Don't move these wires more than necessary; they can break off.

Removing all circuit boards

In the first type of remote control, you'll notice two battery terminals attached to the circuit board that the batteries come in contact with. Take a small flathead screwdriver and pry the battery terminals straight up, as shown in Fig. 37-12. The circuit board now will lift straight up and off.

Another type of circuit board has notches on each side. Some of these notches have plastic latches holding the circuit board in place. Starting at the back of the board, simply place the tip of your fingernail underneath the circuit board while prying it up. At the same time, use your other hand to push the latch back, releasing the circuit board (Fig. 37-13). Follow the same procedure for each latch. After releasing all the latches, the circuit board will lift straight up and off.

For circuit boards with one to five small Phillips head mounting screws, remove the screws and lift the board straight up and off. Don't flip the front cover over after removing the circuit board.

Repairing buttons that stick down

If a beverage has been spilled on the remote or if the buttons stick down and it has stopped working, follow this procedure. In older type remotes when the circuit board is removed, you'll find a rubber pad. This pad usually comes off with the circuit board, while the buttons remain in the front cover, as shown in Fig. 37-14. Place a hand on the back of the front cover and turn the cover on its side, so the buttons won't fall out. Make note of the location of any

37-11   The wire attached to the battery compartment.

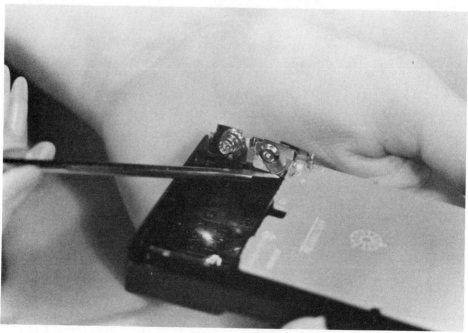

37-12   Prying up the battery terminals.

37-13 Pulling back the latch to release the circuit board.

37-14 The back view of the function buttons.

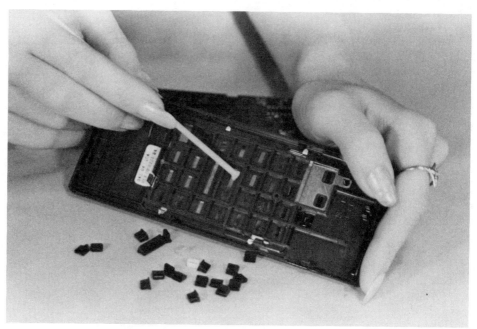

37-15 Cleaning button holes with a glass brush.

colored buttons, so you'll be able to replace the colored buttons back in their proper holes. Flip the cover over and let the buttons fall out onto the working surface.

Spray the front cover and a toothbrush with degreaser. Brush the entire cover, especially down in each hole. Flip the top cover over and clean the other side in the same manner. Then, respray the cover one more time to rinse it off. Take a glass brush and clean each button hole in the front cover, as shown in Fig. 37-15. Spray some degreaser onto your glass brush and scrub each side of each button until it is clean, as shown in Fig. 37-16. Remember to periodically clean the brush with a paper towel. After scrubbing each button clean, make sure to respray them down to rinse them off.

In newer units you find a rubber pad with the buttons incorporated into the pad. Simply pick the pad up (Fig. 37-17) and follow the same procedure as previously explained for cleaning of the top cover and button holes. Spray and brush the top of the pad where the buttons are until it is clean. Be sure to clean around the edges of each button on the pad.

Repairing nonfunctional buttons

The main reason a function button stops working is a buildup of dirt or a beverage on the contact of each button. When the button is pushed, it fails to make electrical contact and the button doesn't work. To fix this problem, take the pad you just cleaned and turn it over. On the back of the pad are round or rectangular shaped conductors. Take a glass brush and dip it into some cleaning alcohol. Use the brush to clean all the conductors, as shown in Fig. 37-18. After cleaning each contact, reclean the brush with a paper towel. After cleaning all the conductors, spray the entire pad off with degreaser to remove any fibers left by the glass brush.

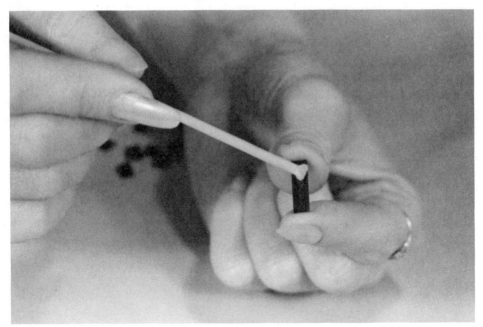

37-16   Cleaning the buttons.

37-17   Removing the pad.

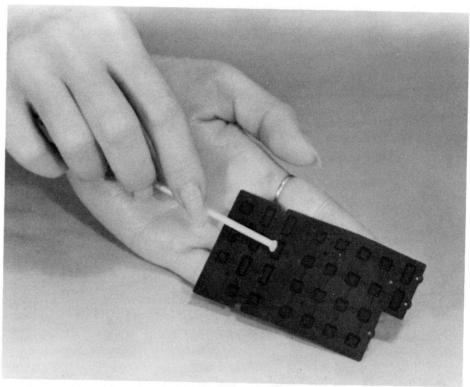

37-18   Cleaning the conductors.

On the back of each circuit board are two types of printed contacts (Fig. 37-19), which look like small and large S's. These printed contacts are what the conductors on the pad come in contact with. To clean these contacts, use a glass brush to clean each contact as you did with the conductors. After cleaning each contact, spray them with degreaser to rinse them off.

In older units where the buttons are separate from the pad, you should find a rubber pad stuck to the bottom of the circuit board. Pull it off, as shown in Fig. 37-20. Clean the conductors on the back of the pad and all the printed contacts on the circuit board as previously explained. This repairs all nonfunctioning buttons.

Repairing dropped remotes

When a remote is dropped the circuit board usually gets broken or cracked. For repairing a cracked or broken circuit board, refer back to chapter 36, "Repairing a VCR that has been dropped."

Reassembling a remote control

In older units, place the top front cover upside down and put all the plastic buttons back into their proper holes. Place the rubber pad over the buttons with the conductors facing up.

37-19 The two types of printed contacts.

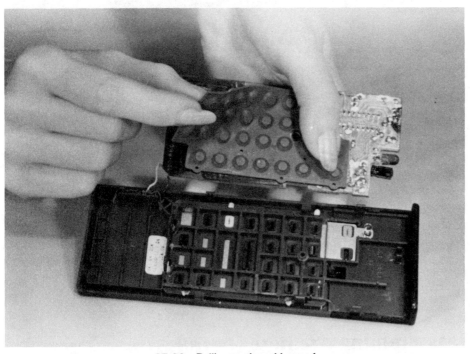

37-20 Pulling up the rubber pad.

37-21   Comparing conductors to the contacts.

Make sure the proper end of the pad is facing towards the front. You can do this by placing the bottom side of the circuit board beside the pad to compare the conductors to their contacts (Fig. 37-21). Place the circuit board so the contacts are facing the pad. Starting at the front of the board, push the board down into its cover. If there are latches on each side of the top cover, push on the circuit board by each latch until the latch slides over the board, as shown in Fig. 37-22. Do each latch in the same manner. Now, turn the remote over and put the buttons in the proper holes. Make sure the buttons are inserted straight.

Some remotes have a red indicator lamp or LED on top of the unit. This lamp is mounted onto a circuit board. When remounting the circuit board, be sure the lamp protrudes through its hole on the front cover. (See Fig. 37-23) The lamp usually has long leads on it. These leads can get bent easily, causing the lamp to be misaligned.

Take the front window and push it down into its proper position. When remounting the back cover to the front cover, start by first sliding in the front end of the cover, as shown in Fig. 37-24. Then, drop the back cover down over the front cover and squeeze them together. You should hear a popping sound coming from each side. This means the latches inside are latched. Now, replace any mounting screws to the back cover. Replace the batteries and the battery cover.

For the type of remote with the buttons incorporated into the pad, place the button side of the pad back into the front cover. Look on the opposite side of the front cover to make sure the buttons are inserted through their proper holes. Now, place the contact side of the circuit board over the pad while placing the LED through its proper hole in the front cover of the unit. If the unit has an indicator lamp, make sure it's properly inserted into its hole as you're fitting the circuit board back into its cover. Align the battery terminals with the

37-22 Remounting the circuit board.

37-23 The base of the indicator lamp.

37-24 Sliding the front end of the cover in first.

grooves and push straight down. Put the back cover up against the front cover and squeeze them together until you hear a popping sound from both sides, indicating that the latches inside have locked into place. Replace the mounting screws, batteries, and battery door cover.

In the last type of remote, line up the circuit board with the pad and replace the circuit board mounting screws. Place both covers back to back, squeezing them together until you hear a popping sound coming from both sides. Then, replace any mounting screws, batteries, and battery door cover.

Review

☐ If the remote doesn't work, replace the batteries. If that doesn't work, then the unit has have to be disassembled for repair.

☐ Look for any screws on the back cover, under the rubber foot pads, or inside the battery compartment.

☐ When removing the back cover, use a small screwdriver to release the latches and/or use a twisting motion to pop the cover off.

☐ Remove the circuit board by taking out the screws, by releasing the latches, or by prying the battery terminals up.

☐ Check the circuit board for small cracks and repair them. Refer to chapter 36.

☐ Check the buttons for sticking. Clean the buttons and the insides of the holes in the front cover.

☐ If you find a function button that doesn't work:

☐ Check the conductors on the rubber pads for dirt or beverages, then clean them.

☐ Check the contacts on the circuit board for dirt or beverages, then clean them.

☐ To reassemble the remote make sure the rubber pads with the conductors are properly aligned to the contacts on the circuit board.

☐ Make sure all the buttons protruding through the front cover are aligned in the proper location and aren't inserted crooked.

☐ Place both covers back to back and squeeze. Listen for a popping sound from each side of the remote.

Glossary

ac cord   The cord that plugs into the 110-volt outlet.

A/C head   A mechanical part containing three sections: the record and playback audio head, the audio eraser head, and the FM head.

alignment screwdriver   A screwdriver made of fiberglass used for adjusting coils.

amperage   The strength of an electric current measured in amperes.

azimuth adjustment   Adjustment on the A/C head mount to peak out the audio signal to its highest peak.

back tension guide   Creates the right amount of torque on the video tape while it is going through the tape path.

bearings   A sleeve which allows the shaft of a dc motor to spin in place.

B+ line   A power line feeding voltage to the circuits.

capacitor   Allows a signal to travel but stops any voltage.

capstan shaft   A powered shaft that pulls the video tape through the tape path.

cassette carriage   Houses the cassette holder and drops the video cassette into the unit.

cassette holder   The part of the cassette carriage that holds the video cassette.

cassette lamp   A light bulb that goes up inside the video cassette after the cassette is inserted into the unit. This lamp is used to activate the tape sensors.

cassette lid   The top cover fastened to the cassette holder.

circuit board   A board holding all electronic components and connecting all components to make a circuit.

clutch arm assembly   The arm that holds the idler wheel and allows the wheel to move to the proper spindle depending on whether you are in fast forward or rewind.

clutch assembly   Applies tension to the idler wheel, allowing the wheel to slip while in play.

component   Any electronic part.

conductors   Anything that allows electricity or electrical impulses to flow through it.

connection   A wire or any two components connected to a terminal point or circuit board.

C-ring   A split spring-loaded ring around a shaft holding a part into place. It has the appearance of the letter C.

dead spot   One of the coils inside a dc motor has opened up. If the motor stops on that spot, the motor will not restart without being spun manually.

drive wheel   A wheel directly driven by a belt or gear off a dc motor.

E-ring   A clip fastened to a notch in a shaft holding a part into place. It has the appearance of the letter E.

electrical circuit   A combination of components making an electronical part function.

electrical contacts   A pair of contacts, one of which carries a signal on it. When the conductor makes contact with them, the signal is allowed to continue to the second contact.

eraser head   Erases all previously recorded information off the video tape.

flywheel   The largest wheel found in the undercarriage. It's always attached to the capstan shaft.

FM adjustment   An adjustment nut used to center the tracking control.

frontload unit   A VCR that loads a video cassette through a front door.

fuse   Opens up to protect a circuit. A warning signal.

gear   A plastic wheel with teeth all the way around it to engage with other connecting parts.

grounding wire   A wire used to ground electronic parts.

height adjustment   A nut adjustment behind the A/C head. It is used to set the audio portion to the proper position on the video tape.

holding bracket   a stationary bracket holding a part in place.

IC   An integrated circuit. It is found on circuit boards

idler arm   An assembly holding the idler wheel, allowing it to make contact with the spindles.

idler wheel   A wheel transfering power from the drive wheel to a spindle.

leaf spring   A spring used to ground a shield or a moving part.

leaf switch   A switch with flexible arms that allow moving parts to activate the switch with ease.

light sensor   Another term for tape sensor, which signals a unit to stop.

load   To load the unit: insert a video cassette and load the video tape onto the tape path.

lock arm   Arms inside of a video cassette that keep the reels from turning when the tape is not in use.

lock nuts   Set screws used to keep a part bonded to another part, such as a pulley fastened to its motor shaft, or to keep a part that has been aligned from turning by itself.

lower video drum   The stationary part of the drum that supports the upper video drum.

microswitch   Switches behind the front panel that activate functions in the unit.

mode   Function that is activated at the time.

mounting bracket   A bracket used to mount a part and hold it in place. This bracket also keeps a part in alignment, while allowing it to be mobile.

mute   Turns the screen black or blue, stopping the picture from reaching the TV monitor.

niche   The center point that can be felt in the knob on a tracking control.

oscillate   To go back and forth and return to the starting point.

pinch roller   The wheel that locks up against the capstan shaft, so the capstan shaft can pull the video tape through the tape path.

pins   Silver or gold connectors with wires attached to the inside of a connector plug.

port-a-pack   A portable VCR model.

pulley   A device that is driven by a belt and transfers its power to another moving part.

reel   A round wheel used to store a video tape.

resistor   An electrical device used in a circuit primarily to provide resistance.

ribbon   A mass of wires inside a flat single plastic coating that is flexible.

roller guide posts   The part of the roller guides that actually come into contact with the video tape. The adjustable part that aligns the video tape to the video heads.

roller guides   The guides pull the video tape onto the tape path and align the video tape to the video drum.

roller guide tracks   Tracks that the roller guides follow to the V-mounts.

rotary cap   The part of a drum motor that drives the video drum.

screen   The image area on a television.

sensor   A device that, when activated, signals the unit to stop.

shaft   A piece of round metal that holds a part in place and allows the part, such as a wheel, roller, or linkage, to rotate on it.

shielded cable   Used for housing video leads to keep out electrical interference.

solenoid   A coil of wire wrapped around a metal shaft and, when activated, pulls the shaft toward the center of the coil to activate other parts.

shield plate   The plate covering the video tape path that keeps outside electrical interference from the video heads.

splicing   To connect two ends of a video tape, a wire, etc.

slide switch   A switch with a post that slides back and forth to activate the switch.

solder point   A joint where two wires are soldered together, a wire is soldered to a terminal, or a pin is soldered to a circuit board.

solder wick   Braided copper wire that soaks up old solder when heated.

sporadically   At variable intervals, not in an even sequence.

supply reel   The reel found in the video cassette that feeds out the video tape onto the tape path.

supply roller guide   The guide that aligns the video tape to the left side of the video drum.

supply spindle   Inserts into the left side of the video cassette. It feeds out the video tape when the unit is in play and pulls the tape in for rewind.

sync   An electronic pulse used to stabilize a picture.

take-up reel   The reel inside the video cassette that takes up the video tape after the video tape has gone through the tape path.

take-up spindle   Inserts into the right side of a video cassette and pulls the video tape into the cassette when in play or fast forward.

tape guides   These guides keep the video tape aligned on the tape path.

tape path   All the components that come into contact with the video tape.

tape sensors   The sensors that signal the unit to stop when the video tape reaches either end.

tension roller   A roller that takes out the vibration or shutter from the video tape.

tilt adjustment   Located on the rear of the A/C head mount and is used to tilt the A/C head back and forth.

topload unit   A VCR that you insert the video cassette into through an opening in the top of the unit.

torque   The pressure applied on a video tape while in play.

tracks   Found on a circuit board, allowing the signals or voltage to travel from one part to another.

transistor   A device that amplifies a signal.

transport   What the cassette carriage and all moving parts are mounted on.

TV monitor   A television set used to diagnose problems in a VCR.

unit   Another description for a VCR.

unload   To unload a video tape from the tape path, to push Eject.

upper video drum   The part of the video drum that spins and contains the video heads.

video amplifier   A circuit that amplifies the video signal or image.

video cassette door   The door on the front of the video cassette that protects the video tape.

video drum   Holds video heads in position and rotates the video heads to pick up the video signal off the video tape.

video heads   The device that picks up the image part of the signal off the video tape.

viewport   The porthole in the video head that allows the image of the video signal to enter the head.

V-mounts   What the roller guides lock into when in the loaded position.

voltage   An electromotive force, or difference in electrical potential, expressed in volts.

worm gear   A shaft having the appearance of a barber pole. It is designed to link with compatible gears, either receiving or imparting motion.

Index